The FOA Reference Guide To
Fiber Optic Testing

Study Guide To FOA Certification

Jim Hayes

I0463293

The Fiber Optic Association, Inc.
The Professional Society Of Fiber Optics
www.foa.org

The FOA Reference Guide To Fiber Optic Testing

And Study Guide To FOA Certification

The Fiber Optic Association, Inc.

1119 S. Mission Road #355, Fallbrook, CA 92028
Telephone: 1-760-451-3655 Fax 1-781-207-2421
Email: info@foa.org   http://www.foa.org

The FOA logo, CFOT® and Fiber U® are registered trademarks of The Fiber Optic Association, Inc., US Patent and Trademark Office Reg. No. 3,572,190.

ISBN. 1-5442-8965-0

Table of Contents

Chapter 9
Optical Time Domain Reflectometer (OTDR) Testing 157

Chapter 10
Fiber Characterization..207

Chapter 11
Reflectance Testing...231

Chapter 12
Testing A Passive Optical Network (PON) 245

Chapter 13
Testing And Troubleshooting Checklists 257

Preface

The purpose of this book is to provide a reference guide to those involved with the testing of fiber optic cable plants and networks or those teaching the personnel who will do this work. This book is also the reference guide for FOA CFOS/T Design Specialist certification.

Many in the fiber optic industry say that fiber optic testing is the biggest problem faced by manufacturers, installers and network operators. The FOA agrees with this assessment since testing questions are the most often missed questions on FOA certification tests.

Testing is needed to verify components and the quality of installations. Testing is needed to troubleshoot networks. The whole of fiber optics depends on testing, yet it seems to be the least understood topic. This book has been written to provide a reference guide to fiber optic testing that will be understandable and provide a definitive reference to the industry.

It is assumed that the reader is familiar with basic fiber optic technology used for premises and outside plant networks at the level of an FOA CFOT. If not, you should begin by studying general fiber optics using Chapter 2 - Jargon in this book, the other FOA textbooks, FOA Online Guide (www.foaguide.org) or take the Basic Fiber Optics self-study course at Fiber U (www.fiberu.org).

The Fiber Optic Association, Inc., the nonprofit professional society of fiber optics, has become one of the principal sources of technical information, training curriculum and certifications for the cabling industry.

The FOA is focused on the education and certification of technicians, not selling products. We think it's important that every tech knows as much as possible about fiber optic technology so they can communicate with others regarding the technology and deal successfully with design and installation in all types of communications applications.

The FOA created its Online Reference Guide (www.foaguide.org) to provide a more up-to-date and unbiased reference for those seeking information on cabling and fiber optic technology, components, applications and installation. Its success confirms the assumption that many users prefer the Internet for technical information. Much of the information in this book comes from the FOA Guide.

With this book, we address the needs for those who prefer printed books or who must have them to meet academic requirements. However, the production of this book is done by "publishing on demand," where the book is not printed until ordered, and only after accessing the latest version electronically. Thus this edition meets the needs of those who prefer printed references without burdening them with trying to determine what material is already obsolete.

You may be interested in the other FOA textbooks:
The FOA Reference Guide To Fiber Optics (available in English, Spanish and French), ISBN 1-4392-5387-0
The FOA Reference Guide To Outside Plant Fiber Optics, ISBN 1450559670
The FOA Reference Guide To Premises Cabling, ISBN 1450559662
The FOA Reference Guide To Fiber Optic Network Design, ISBN 1-5308-8635-X
The FOA Outside Plant Construction Guide, ISBN 1523925825

For those who want this printed version but also want access to the web for color graphics, self-study programs with automatic self-testing or links to even more technical information, we suggest going to the FOA Online Reference Guide (foaguide.org) website Table of Contents to find the appropriate sections covered in this book. The Guide also has a custom search engine to allow searching the entire Guide for topics of interest.

If you have feedback on the book, feel free to email comments or questions to the FOA at info@foa.org.

A note of appreciation
The material in this book has been contributed and reviewed by a number of FOA instructors whom we wish to thank for their work: Joe Botha, Tom Collins, Duane Clayton, Stephen Cook, Ed Forrest, Ian Gordon Fudge, Frank Ghassemi, Gary Giguere, Bill Graham, Bee Suat Lim, Terry O'Malley, Steve Wolszczak and all the others we've forgotten who helped in creating this book. We'd like to especially thank Terry O'Malley and Steve Wolszczak for their contributions on OTDRs and Duane Clayton for his editorial skills.

Jim Hayes, Writer & Editor

This information is provided by The Fiber Optic Association, Inc. as a benefit to those interested in testing fiber optic communications systems or networks. It is intended to be used as a overview and/or basic guidelines and in no way should be considered to be complete or comprehensive. These guidelines are strictly the opinion of the FOA and the reader is expected to use them as a basis for learning, reference and creating their own documentation, project specifications, etc. The FOA assumes no liability for the use of any of this material. Those working with fiber optics in the classroom, laboratory or field should follow all safety rules carefully.

Chapter 1

Introduction To Fiber Optic Testing

Objectives: From this chapter you should learn:
What is fiber optic testing
Why is fiber optic testing important
What needs to be tested and how it is tested
What is the relevance of industry standards

Introduction

Fiber optic testing is done to verify the performance of a fiber optic component, cable plant or communication system. Testing involves examining components visually to determine their condition and measuring their performance with sophisticated instruments according to standardized test procedures.

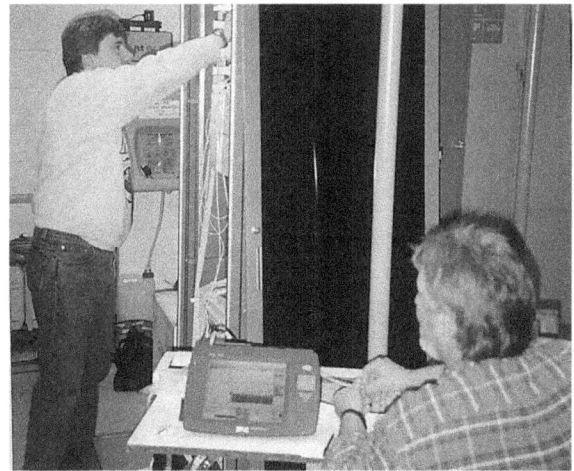

Testing an installed fiber optic cable plant

Testing may be done during research and development, engineering, manufacturing, installation, operation and troubleshooting and repair. In this book, the focus is primarily on installation, operation and troubleshooting, although much of what is covered here is relevant to every aspect of fiber optics.

Fiber optic manufacturers will tell you the biggest problem they have with installations of their products is testing by the installers. Contractors and installers are often unclear about how to test fiber optics or how to interpret the data they get from fiber optic test equipment. Network owners likewise are often given data by their contractors regarding an installation that they do not understand.

We can vouch for that at The FOA. Testing is the subject of most of the help request calls and emails we receive and it is the subject of questions most often missed on the Fiber U certificate of completion tests and FOA certification exams. Testing is also the largest subject area for the FOA Online Guide website. Beyond the FOA, testing is the subject of most industry standards from around the world.

Why is fiber optic testing such a big issue? Testing can be complicated. Certainly fiber optic test instruments have gotten more complicated and often more time is spent learning how to use an instrument than how it makes measurements or how to interpret the data it produces. Many of those instruments are so automated that users often know little about what is involved in a test.

This book is written to provide the knowledge of what is involved in a test and how to interpret the data.

What Needs Testing

Fiber optic testing involves several types of tests that provide different kinds of data. Much of the testing can be summarized in these questions:
1. Do the connectors on fiber optic cables look OK?
2. Are the end-to-end connections correct?
3. How well does the fiber optic cable plant transmit light?
4. If something is wrong, what's going on with the fiber and where is the problem?
5. If the system is connected, is the optical power at the transmitter and receiver at the correct level?
6. How accurate are these measurements?

Answering each of those questions requires different test equipment and different test procedures. In this book we will examine those questions (and quite a few more) and see what they mean in much detail. We will also look at that last topic, the accuracy of the measurements in much more detail than you have probably seen before.

Understanding fiber optic testing and the accuracy of the measurements is the key to successful testing and troubleshooting. And it is the focus of this book.

Fiber Optic Cable Plant Testing

Cable plant testing is the most common task for the fiber optic tech involved in testing. The process includes five basic steps to evaluate the condition of the cable plant.

Do The Connectors On A Fiber Optic Cable Look OK?

Connectors cause most problems with fiber optics, so they need to be tested first. Connector problems cause high loss in a fiber link that can affect transmission of data. Most connector problems are caused by dirt on the end of the connector or poor connector installation and polishing. The first and most important step in testing connectors is to inspect them visually. This is done with a special fiber optic inspection microscope and you must look for dirt or other contamination, scratches or cracks that can affect the transmission of light.

Microscope view of connector showing dirt, oil and scratches.

Of course if the connector is dirty, you must clean it. Cleaning connectors is actually a complex topic itself (see Chapter 5) that is related to testing – you simply cannot test with dirty or damaged connectors. You can't just wipe a connector off with a tissue or on your shirt and plug it back in. To really get a connector clean, you need some of the special products made for cleaning them. And after you clean the connector you need to do a visual inspection again to ensure the connector was cleaned properly.

Are The End-To-End Connections Correct?

Often when a fiber optic link being connected doesn't work, the connections between transmitter and receiver are not correct. If there are just two fibers, it's easy to switch them and try again,. If you are trying to find the two correct fibers in a multifiber cable, there should be documentation that refers to the cable color codes. There are also visual tracing testers that allow you to trace fibers for many kilometers to find the correct fibers. Those visual tracers will be quite necessary if you are working on a large fiber count cable, as is often the case.

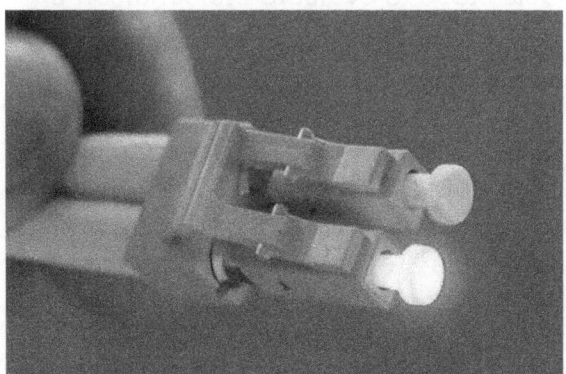

Tracing fibers with a visual tracer

How Well Does The Fiber Transmit Light?
The light transmission of an optical fiber is tested with instruments that work like the transmission system that will be connected with the fiber. For a transmitter, we use a test light source with a LED or laser source similar to the transmitter. For a receiver, we use an optical power meter. The combination of light source and power meter is called an optical loss test set or OLTS. Instead of patchcords to connect the light source and power meter to the fiber we want to test, we use high quality reference cables attached to the light source and meter.

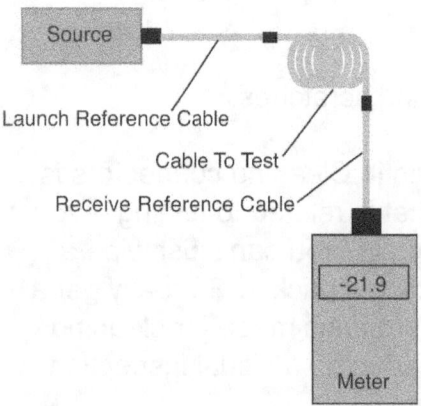

A test source, power meter and reference cables used for testing a cable plant

The light source, power meter and reference cables are calibrated before testing to set a "0 dB" loss reference. Then the fiber to test is connected between the two reference cables and the meter will read the loss in dB. That loss is compared to one or both of two values, either the estimated loss of the fiber based on average component losses (we call that a "loss budget") or the maximum dB loss the system transmitter and receiver can tolerate (the "power budget.")

This test is called an "insertion" loss test because we insert the fiber to be tested between the reference cables attached to the light source and power meter. Every fiber needs to be tested this way and every international standard calls for this as the primary test for all fibers. The loss for every fiber is recorded in the documentation for the cable. Chapter 8 covers insertion loss in detail.

Visual Fault Location

Sometimes, even if the fiber has been tested for insertion loss, we need to know more. Insertion loss only tells us the total loss of all parts of the fiber including any splices, connections or losses caused by installation problems like bends in the fiber that are tighter than allowable which can cause stress loss.

In short cables, a visual fault locator (VFL) can be extremely helpful. A VFL uses a high power visible red laser coupled into the fiber to allow identifying fibers and tracing cables using the visible light transmitted through the fiber. In simplex cables used for patchcords or tight-buffered fiber you can see light lost if the fiber or cable is bent too tightly. You can also see the light lost around bad connectors or splices if they ware within a few kilometers of the VFL.

Taking A "Snapshot" Of The Fiber

Another test, called an OTDR test for the name of the instrument used, an Optical Time Domain Reflectometer, uses scattering from the fiber to provide a picture of the fiber along its length. The OTDR shows individual fiber, splice, connector and stress losses and their position along the length of fiber. OTDRs are very effective for troubleshooting as well as testing. Chapter 9 covers OTDR testing in detail.

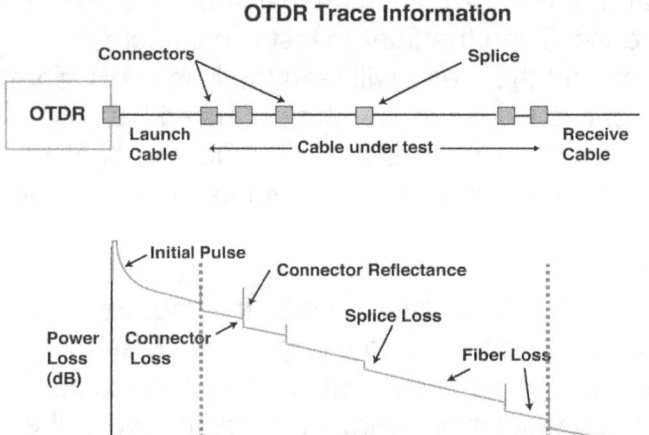

OTDR trace with markers set to measure distance

Sometimes very long outside plant singlemode cables also need testing for dispersion that limits the bandwidth of signals that can be transmitted. There are several types of dispersion that can be tested and the extremely high bandwidth of fiber means these tests are complicated and expensive.

Testing dispersion is common only for long outside plant (OSP) singlemode cable plants on which very high speed networks will be operating. New cable plants are generally tested for chromatic dispersion (CD) and polarization mode dispersion (PMD) as they are installed. Older cable plants are tested to determine their capability for network upgrades. Details on testing fiber for dispersion, called fiber characterization, is in Chapter 10.

Premises Cables Or Outside Plant Cables?

Premises (indoor) cables used in LANs (local area networks for computers), closed circuit TV (security cameras), wireless distributed antenna systems (DAS) or building management systems are generally short cables that have low loss, with loss dominated by connector loss. These cables sometimes run very fast networks that have very low power budgets (~2dB) so testing is done carefully with a light source and power meter to minimize test uncertainty. OTDRs are generally not as useful for testing premises cables because they lack distance resolution adequate for short cables and reflectance from connectors overwhelm the OTDR receiver. Premises cables rarely have splices to test either, one of the main reasons to use an OTDR.

Long outside plant (OSP) cables are also tested with a light source and power meter to get loss. But OSP cables are also almost always tested with an OTDR. The OTDR can verify splice loss, although that often requires testing in both directions to get valid data. The OTDR can also find stress points induced in installation. One big use for OTDRs is to provide documentation for the future, in case there is a problem like a dig-up that requires restoration, when having data on what the cable plant looked like at installation will be invaluable in tracing and finding problems.

Testing Communications Networks

Fiber optic communications network communicate over fiber optic links. A link consists of a transmitter coupled into a cable plant that connects to a receiver on the far end. Full duplex links will usually consist of two fibers transmitting in opposite directions, although some networks like fiber to the home use bidirectional transmission over a single fiber.

Duplex fiber optic link

The first requirement for proper transmission is that there be the proper amount of power at the receiver – not too little power which causes problems with poor signal to noise nor too much which overloads the receiver. Measuring optical power allows testing the transmission system for proper optical connections. Power measurement is covered in Chapter 7.

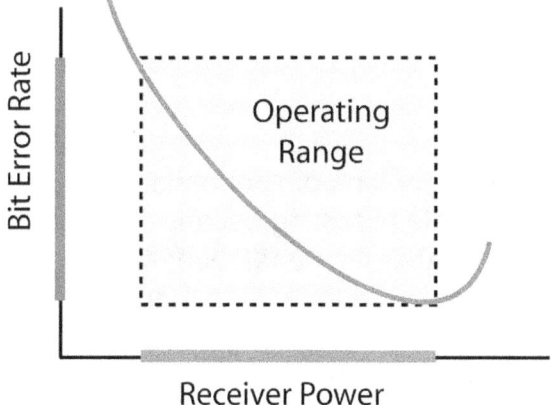

Bit error rate is a function of receiver optical power

The amount of power at the receiver is determined by the amount of power coupled by the transmitter into the fiber optic cable plant that is then reduced by the attenuation of the cable plant. During the design stage, the link power budget and fiber optic cable loss budget are calculated to ensure the design can accommodate the fiber optic links chosen. Chapter 6 covers loss and power budgets.

To test a fiber optic link, you turn on the transmitter then test the power at the receiver end of the cable plant. You compare that power measurement to the manufacturer's specification for the receiver. You want the power level to be in the correct range for that receiver.

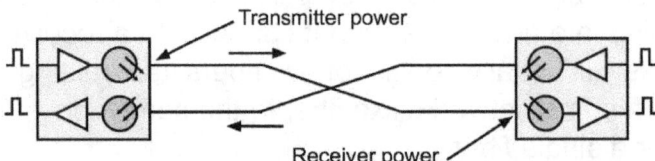

Testing optical power in a fiber optic link

If the receiver power is too high, you add a fiber optic attenuator at the receiver to reduce the power to the proper level. If the power is too low, you go to the transmitter and measure the power there to see if it is within the manufacturer's specification.

The power at the transceiver may be correct and the cable plant loss is too high, so the power at the receiver is too low. This means you need to reduce the loss in the cable plant or find a higher power transmitter. If the cable plant loss is proper but the transmitter power is too high or too low, you will need to find a transmitter with proper power level.

Industry Standards

There are industry standards for practically every kind of fiber optic test. In fact the majority of fiber optic standards relate to testing because if the importance of testing to proper network installation and operation. Many standards relate to component testing, especially under environmental stress like temperature, moisture, altitude, etc. There are also standards for testing cable plants and links that include guidelines to ensure measurements are as accurate as possible.

Chapter 3 covers the principle of standards and includes is a list of US and international standards. These standards change continuously as old ones are dropped, current ones are updated and new ones are added. Check with the appropriate standards groups to get the most up to date information.

Measurement Accuracy Or Uncertainty

If you give a tech an instrument, they rarely think about what is the measurement uncertainty of the tests made with that instrument. Especially with digital instruments, the display is assumed to be accurate. In fact, every test and every instrument has some uncertainty and it is generally much higher than the resolution of the digital readout on the instrument.

Understanding metrology, the science of measurement, is important in understanding how to interpret the data acquired in a test. In the final chapter of this book (Chapter 14), we will discus the metrology of fiber optics and provide guidelines on how to interpret the uncertainty of fiber optic measurements.

Training Technicians In Testing

Fiber optic testing involves more than owning the right equipment and knowing how to operate it. Test technicians need to know what the tests are testing, how the instrument makes those tests and how to troubleshoot problems. Those are all topics that will be covered in later chapters in this book. They are also covered in FOA's training curriculum taught at our approved schools around the world and available for self study free on Fiber U online.

For those interested in studying on their own, this book covers the information that you should know and chapter quizzes allow you to test your knowledge.

Chapter Quiz

1. Most of the problems with fiber optics are caused by _____.
   A. Bad fibers
   B. Poor splices
   C. Bad or dirty connectors
   D. Transmission equipment

2. To test the fiber optic cable plant in a manner in similar to how transmission equipment uses the fiber, you need a _____.
   A. Visual inspection microscope
   B. Test source and power meter
   C. OTDR
   D. Visual fault locator

3. What instrument takes a snapshot of the fiber in a fiber optic cable plant?
    A. Visual inspection microscope
    B. Visual fault locator
    C. Test source and power meter
    D. OTDR

4. When testing a fiber optic transmission system, you should first test
    _____.
    A. Cable plant loss
    B. Transmitter power
    C. Receiver power
    D. Connector and splice loss

5. The majority of fiber optic standards cover _____.
    A. Component specifications
    B. System specifications
    C. Testing
    D. Troubleshooting

Chapter 2

Fiber Optic Testing Jargon

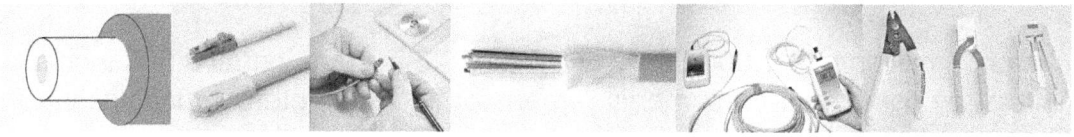

Objectives: From this chapter you should learn:
The language of fiber optics
Systems of measurements used in fiber optics
Specialized fiber optic terms

Introduction

The key to understanding any technology is understanding the language of the technology – the jargon. We've started this book with an overview of fiber jargon to introduce you to the language of fiber optics and help you understand what you will be reading in the book. While we try to use the most common terminology, some particular applications for optical fiber have their own specialized terms, and when possible, we will try to include those terms also. We suggest you read this section first to help your understanding of the rest of the book and refer back to it when you encounter a term that you do not recognize. You can also use the Definitions in the Appendix of this book or the FOA Online Reference Guide for more explanations.

Table of Contents: The FOA Reference Guide To Fiber Optics

What Is Fiber Optics?

Fiber optic communications means sending signals from one location to another in the form of modulated light guided through hair-thin fibers of glass or plastic. These signals can be either analog or digital and transmit voice, data or video. Fiber can transport more information longer distances in less time than any copper wire or wireless method. It's powerful and very fast - offering more bandwidth and distance capability than any other form of communication!

The Metric System

Fiber optics, as an international technology, utilizes the metric system as the standard form of measurement. Several of the more common terms used are:

Meter: 3.28 feet, 39.37 inches. Fiber optic cable lengths are generally expressed in meters or kilometers.

Kilometer: 1000 meters / 3,281 feet / 0.62 miles.

Micron: 1 millionth (1/1,000,000th) of a meter. 25 microns equal 0.001 inch. This is the common term of measurement for fiber diameters. Most glass fibers are 125 microns in outside diameter.

Nanometer: One billionth of one meter. This term is commonly used in the fiber optics industry to express wavelength of transmitted light, e.g. 850 or 1300 nm.

Fiber

Optical Fiber: Thin strands of highly transparent glass or plastic that guide light, used to transmit communications signals.

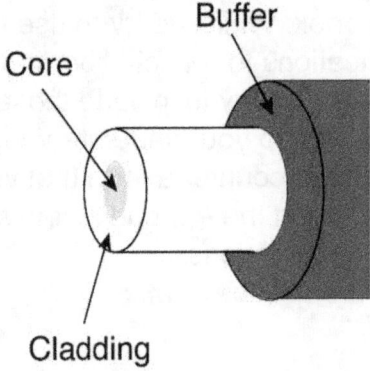

Core: The center of the fiber where the light is transmitted.

Cladding: The outside optical layer of the fiber that traps the light in the core and guides it along - even through curves.

Buffer coating or primary buffer coating: A hard plastic coating on the outside of the fiber that protects the glass from moisture or physical damage. The buffer is what one strips off the fiber for termination or splicing.

Mode: A single "electromagnetic field pattern" or ray of light that travels in fiber.

Multimode fiber: has a larger core (almost always 50 or 62.5 microns - a micron is one millionth of a meter) that transmits many modes or rays of light.

Multimode fiber is used with laser or LED sources at wavelengths of 850 and 1300 nm for short distance applications like LANs or security cameras.

Singlemode fiber: has a much smaller core than multimode fiber, only about 8-9 microns, so it only transmits one mode. It can go very long distances at very high speeds. Singlemode is used for telephony (long distance, metropolitan and fiber to the home) and CATV usually with laser sources at 1310 to 1550 nm or over a broader range for wavelength division multiplexing.

Fiber identification: Fibers are identified by their core and cladding diameters expressed in microns (one millionth of a meter), e.g. 50/125 micron multimode fiber or 9/125 micron singlemode fiber. Most multimode and singlemode fibers have an outside diameter of 125 microns - about 0.005 - 5 thousandths of an inch - just slightly larger than a human hair. International standards for fibers call out detailed specifications that also include bandwidth capability or other special characteristics.

Plastic optical fiber (POF): is a large core (usually 1mm) multimode fiber that can be used for short, low speed networks. POF is used in consumer HiFi and as part of a standard for vehicle communication systems called MOST.

Fiber Optic Cable

Cable: Cable provides protection to the fibers from stress during installation and from the environment once it is installed. Cables may contain from only one to hundreds of fibers inside. Cables come in three varieties: tight buffer with a thick plastic coating on the fibers for protection, used mainly indoors, loose-tube, where fibers with only a primary buffer coating are inside plastic tubes, and ribbon, where fibers are made into ribbons to allow small cables with the largest numbers of fibers.

Many international standards refer to simplex cables as "cords".

Jacket: The tough outer covering on the cable. Cables installed inside buildings must meet fire codes by using special jacketing materials.

Strength members: Aramid fibers (Kevlar is the Dupont trade name) used as strength members in the cable to allow pulling the cable or suspending it for aerial installation. The term is also used for the steel or fiberglass rod in the center of some cables used to stiffen it to prevent kinking as well as provide pulling strength.

Armor: Prevents crushing and discourages rodents from damaging cable by chewing through it. Some cable called armored also includes layers of strengthening wires for use in extreme environments such as encountered by submarine cables.

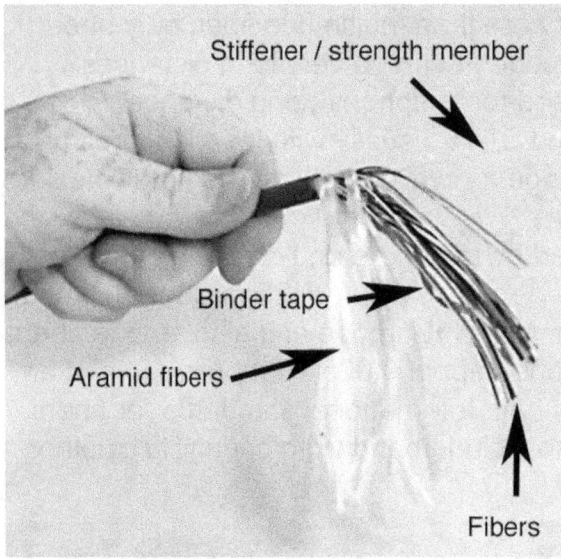

Sheath: A term used for the combination of the jacket, armor and any other elements used to protect the fibers in a cable.

Fiber Optic Installations

Outside Plant Installations
Outside plant installations fall into four general categories, depending on the placement of the cable. Each requires cable types chosen for the installation and specialized equipment for placement.

Underground: Cables placed underground in conduit, often inside innerduct pulled in the conduit. Cables can also be blown into duct lines installed by trenching or plowing.

Direct Buried: Cable placed underground without conduit, placed in trenches, plowed into the ground or installed by directional boring.

Aerial: Cable placed above ground on utility poles.

Submarine: Cables placed underwater, including those in shallow water such as lakes or rivers as well as those used for ocean crossings.

Premises Installations
Premises (indoor) installations are installed in cable trays, on J-hooks, in ducts or conduit inside walls, above ceilings, etc.

Premises (indoor) installations generally use tight buffered cables with jackets that are rated for flame retardance for safety.

Termination and Splicing

Connector: A non-permanent device for connecting two fibers in a non-permanent joint or connect fibers to equipment. Connectors are expected to be disconnected occasionally for testing or rerouting fibers.

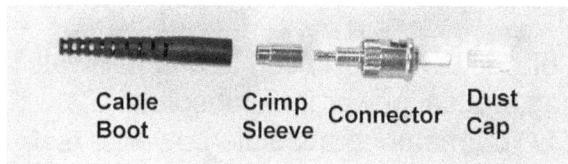

Splice: a permanent joint between two fibers primarily used to *concatenate* (join) long fibers in outside plant installations and attach pigtails to terminate them.

Mechanical Splice: A splice where the fibers are aligned created by mechanical means.

Fusion Splice: A splice created by welding or fusing two fibers together.

Fusion Splicer: An instrument that splices fibers by fusing or welding them, typically by electrical arc.

Hardware: Terminations and Splices require hardware for protection and management: patch panels, splice closures, etc.

Fiber Performance Specifications

Attenuation: The reduction in optical power as it passes along a fiber, usually expressed in decibels (dB). For fibers, we talk about attenuation coefficient or attenuation per unit length, in dB/km. See optical loss

Bandwidth: The range of signal frequencies or bit rate within which a fiber optic component, link or network will operate.

Decibels (dB): A unit of measurement of optical power that indicates relative power. dB is a logarithmic scale where 10 dB equals a factor of 10 times. For example, 3 dB is a factor or two, 10 dB a factor of ten. Negative dB indicates loss, so -10 dB means a reduction in power by 10 times, -20 dB means another 10 times or 100 times overall, -30 means another 10 times or 1000 times overall and so on.

dB: Optical power referenced an arbitrary zero level, used to measure loss.

dBm: Optical power referenced to 1 milliwatt, used to measure absolute optical power from transmitters or at receivers. See optical power.

Loss, optical Loss: The amount of optical power lost as light is transmitted through fiber, splices, couplers, etc, expressed in "dB."

Optical Power: is measured in "dBm", or decibels referenced to one milliwatt of power. While loss is a relative reading, optical power is an absolute measurement, referenced to standards. You measure absolute power to test transmitters or receivers and relative power in "dB" to test loss.

Dispersion: Pulse spreading caused by modes in multimode fiber (modal dispersion), the difference in speed of light of different wavelengths (CD or chromatic dispersion in multimode or singlemode fiber) or polarization (PMD or polarization mode dispersion in singlemode)

Scattering: The change of direction of light after striking small particles that causes the majority of loss in optical fibers and is used to make measurements by an OTDR

Wavelength: A term for the color of light, usually expressed in nanometers (nm) or microns (m). Fiber is used mostly in the infrared region where attenuation is lowest but infrared light is invisible to the human eye. Most fiber specifications (attenuation, dispersion) are dependent on wavelength.

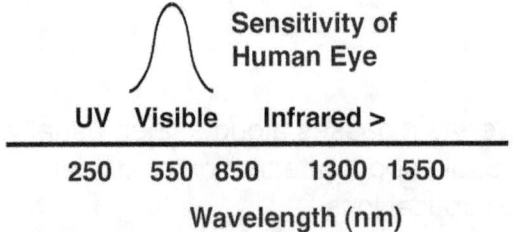

Tools

Jacket Slitter or Stripper: A cutter for removing the heavy outside jacket of cables

Fiber Stripper: A precise stripper used to remove the buffer coating of the fiber itself for termination. There at three types in common use, called by their trade names (from left): "Miller Stripper", "No-Nik" and "Micro Strip."

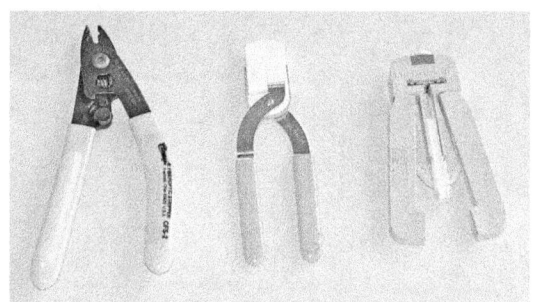

Cleaver: A tool that precisely "breaks" the fiber to produce a flat end for polishing or splicing.

Scribe: A hard, sharp tool that scratches the fiber to allow cleaving.

Polishing Puck: for connectors that require polishing, the puck holds the connector in proper alignment to the polishing film.

Polishing Film: Fine grit film used to polish the end of the connector ferrule.

Crimper: A tool that crimps the connector to the aramid fibers in the cable to add mechanical strength.

Fusion Splicer: An instrument that welds two fibers together into a permanent joint.

Fiber Optic Test Equipment

Optical Power Meter: An instrument that measures optical power from the end of a fiber.

Test (Light) Source: an instrument that uses a laser or LED to couple an optical signal into fiber for testing loss of the fiber or cable

Optical Loss Test Set (OLTS): A measurement instrument that includes both a optical power meter and light source used for measuring insertion loss of installed cable plants or individual cables. The combination of meter and source, either separate instruments or combined, is sometimes also called light source and power meter (LSPM.)

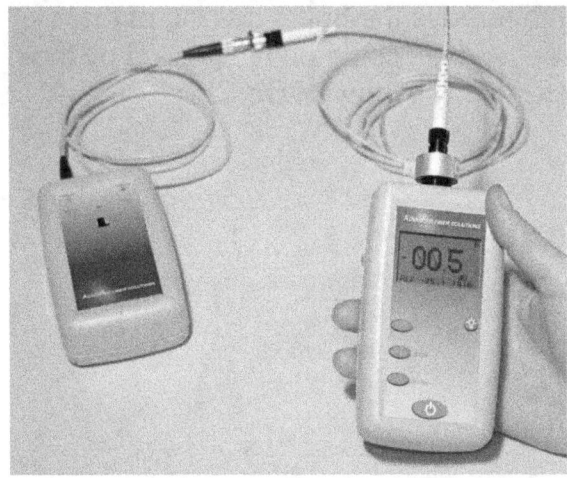

Reference Test Cables: short, single fiber cables with connectors on both ends, used to test unknown cables. A launch cable is attached to the source and used to set the reference power for loss measurements and a receive cable is attached to the power meter.

Mating Adapter: also called splice bushing or couplers, allow two cables with connectors to mate.

Fiber Tracer: An visible light source (LED or flashlight) that allows visual checking of continuity and tracing for correct connections such as duplex connector polarity

Visual Fault Locator (VFL): A high-powered visible laser light source that allows continuity testing, fiber tracing and location of faults near the end of the cable.

Inspection Microscope: used to inspect the end surface of a connector for faults such as scratches, polish or dirt.

Optical Time Domain Reflectometer (OTDR): An instrument that uses backscattered light to take a snapshot of an optical fiber which can be used to measure fiber length, splice loss, fiber attenuation and for fault location in optical fiber from only one end of the cable.

Specialized Testers: Long distance networks may need testing for chromatic dispersion (CD) and polarization mode dispersion (PMD). Systems using wavelength-division multiplexing may need testing for spectral attenuation. Each performance factor has a specialized tester for that specification.

Chapter Exercises

Review manufacturer's websites, catalogs or datasheets to see what fiber optic products are available. See if they use different terminology and how much of it is based on trade names.

Chapter Quiz

True/False
Indicate whether the statement is true or false.

1. Optical fibers can transmit either voice, data or video using either analog or digital signals.
> True
> False

2. Singlemode fiber has a smaller core than multimode fiber.
> True
> False

Multiple Choice
Identify the choice that best completes the statement or answers the question.

3. In an optical fiber, the light is transmitted through the _____.
> A. Core
> B. Cladding
> C. Buffer
> D. Jacket

4. The diameter of an optical fiber is traditionally measured in _____.
> A. Meters
> B. Millimeters
> C. Microns (micrometers)
> D. Nanometers

5. Rays of light transmitted in multimode fiber are called _____.
> A. Reflections
> B. Refractions
> C. Waves
> D. Modes

6. Loss of a fiber or any fiber in a cable is measured in _____.
   A. dB
   B. dBm
   C. milliwatts

7. 10 dB corresponds to a factor of _____ in power.
   A. 2
   B. 5
   C. 10
   D. 100

8. A fiber stripper removes the _____ of the fiber.
   A. Core
   B. Cladding
   C. Buffer coating

9. The _____ protects the fiber from environmental harm.
   A. Primary buffer coating
   B. Aramid fiber strength members
   C. Jacket
   D. All of the above

10. Which fiber optic test instrument uses backscattered light for measurements?
    A. OLTS
    B. OTDR
    C. VFL
    D. Tracer

11. The wavelength of light used for most fiber optic systems is in the _____ region and _____ to the human eye.
    A. ultraviolet, invisible
    B. solar, visible
    C. infrared, invisible

Chapter 3

Fiber Optic Testing Standards

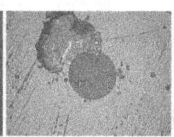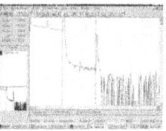

Objectives: From this chapter you should learn:
What is a standard
How standards for fiber optic testing are created
What should be included in a standard
Why the FOA creates its own standards

What Is A Standard?

Here is one definition by the world's standards organization, the ISO (Interational Standards Organization.) ISO/IEC Guide 2:1996, definition 3.2 defines a standard as:
A document established by consensus and approved by a recognized body that provides for common and repeated use, rules, guidelines or characteristics for activities or their results, aimed at the achievement of the optimum degree of order in a given context.
Standards provide rules that help bring order to processes and products.

Standards have existed as long as commerce has. Without standards it would be impossible to say how big something is (length standards in feet or meters) or much it weighs (weight in pounds or mass in kilograms). Time needs a standard second to define the length of time one works on a job. And so on. Throughout history we have created standards that allow buyer and seller to have a common language for commerce. That continues today in our high tech world.

No application in the communications industry could work without industry standards. Any standard's main goal is to create uniform specifications for products that ensure interoperability among various manufacturers' products. Standards start at the component level that cover specifications for connectors and cables, for example, making them intermateable and procedures on how to test them. Fibers are standardized so they can be connected or spliced and used by standard communications systems.

Standards at the system level cover signal bitrates, frequencies and amplitudes, protocols, data encoding, packet length, timing, error correction and of course, the cable plant or radio frequencies they operate over. There many other more technical factors that are needed to guarantee that systems can talk to each other. Communications systems like cell phones, Ethernet and WiFi rely on industry standards, as does the cabling that connects them.

New standards are continually being written and old standards are being rewritten to keep up with technology changes. Most standards are on a 5-year update cycle but that is somewhat arbitrary. Some standards have lasted for decades, like the basic fiber optic testing standards, while some are outdated practically as soon as they are approved.

Covering standards in textbooks and training programs that cover cabling is difficult. Standards change continuously, with the written, approved versions often lagging current product technology by months or even years. Instructors and authors must use care as they refer to standards, covering the scope of the current and expected future versions. The best method of understanding standards is to depend on the manufacturers who write the standards and make products according to them. Their continual involvement with the standards process and dependence on them for product sales ensures they have the most up to date information.

Standards For Fiber Optic Testing

Standards for testing fiber optic components, cable plants and communications systems should reflect the way testing is done and how the results are interpreted. Basically the process can be divided into six steps.

1. What needs to be tested?
2. What are the procedures for making the test?
3. What equipment is needed to perform the tests?
4. What are the options required to implement the tests under various conditions?
5. What are the sources of error in the measurements?
6. What documentation is required for the tests?

The order of these steps is important, since it is the sequence of how one solves the problem of determining performance values for the component, subsystem or system under test and established the validity and precision of the measurement. Once one describes the test methodology itself, it is appropriate to describe the contributions to the uncertainty of the measurement and ways to reduce that uncertainty.

Unfortunately, few standards follow these logical steps. Some have evolved over decades of use and updating to become a patchwork of information. Others have been written or edited to emphasize some aspect of the testing process. Many are hard to understand so the FOA has undertaken a project to create standards that interpret the basic standards used in fiber optics.

Test equipment manufacturers like to talk about "certifying" the cable plant. What they mean is testing the cable plant to determine if the cable plant meets the specifications for the cable plant in the standards, something copied from the unshielded twisted pair copper cable standards (what most people call "Cat 5"). "certifying fiber optic cable plants are much more difficult, since the standards are not as restrictive. What most people consider certifying is comparing the cable plant to the loss budget created at the cable plant design phase, something we will discuss further in Chapter 6 on loss budgets.

The FOA Standards

Since the TIA and ISO/IEC standards were written by manufacturers for manufacturers, they often are not relevant for cable plant designers, contractors, installers or users, the people who are the FOA constituency. The FOA has been involved in these standards committees for decades, but finally decided to write our own standards for our audience - cable plant designers, contractors, installers and users. FOA standards are written to be easily understood and applied, as well as relevant to the applications, and follow other industry standards for the components and communications systems which communicate over these cable plants. See FOA Standards

Our favorite quotes on standards:

"Standards are mutually agreeable specifications for product development." (Former head of TIA 802.3 Ethernet standards committee)

"The wonderful thing about standards is we have so many to choose from." (Bob Metcalfe, co inventor of Ethernet)

"Why do we have international standards? Because those who control the standards control the marketplace." (Massachusetts Port Authority)

"We don't write standards for installers or users, we write standards for manufacturers." (Former head of TIA TR-42 cabling standards committee)

The FOA has tried to gather together information on standards for both components and networks using fiber optics and premises cabling. In keeping with what we say above, we expect these references to be outdated and urge you to contact manufacturers for the latest information.

The FOA personnel have been voting members of TIA TR-42 and prior committees dealing exclusively with fiber optics since the early 1980s and attend most meetings to monitor activities and contribute our knowledge from many years in fiber optics.

International Standards For Fiber Optics

The following is a list of international standards that cover various topics in fiber optics. Remember that all of these standards are revised regularly and therefore such a list is guaranteed to be out of date. Go to the relevant standards bodies to get the latest information.

TIA Fiber Optic Standards
These standards are under continual updating so it's unlikely this list is ever fully up to date.

A full catalog of TIA standards is at http://www.tiaonline.org/

TIA Fiber Optic Test Procedures (FOTPs)
(These are commonly known as "FOTPs" but are officially called "TIA-455-x, e.g. TIA-455-34 is FOTP-34. As they change continually, this list should be considered for reference purposes only - see the TIA for a catalog of the latest versions.)
FOTP-1 - Cable Flexing for Fiber Optic Interconnecting Devices (ANSI/TIA/TIA-455- 1-B-98)
FOTP-2 - Impact Test Measurements for Fiber Optic Devices (ANSI/TIA/TIA-455-2-C- 98)
FOTP-4 - Fiber Optic Component Temperature Life Test
FOTP-5 - Humidity Test Procedure for Fiber Optic Components
FOTP-6 Cable Retention Test Procedure for Fiber Optic Interconnecting Devices
FOTP-8 - Measurement of Splice or Connector Loss and Reflectance Using an OTDR
FOTP-10 - Procedure for Measuring the Amount of Extractable Material in Coatings Applied to Optical Fibers (withdrawn April, 1996)
FOTP-11 - Vibration Test Procedure for Fiber Optic Components and Cables
FOTP-12 Fluid Immersion Test for Fiber Optic Components
FOTP-13 Visual and Mechanical Inspection of Fibers, Cables, Connectors,

and Other Devices

FOTP-14 - Fiber Optic Shock Test (Specified Pulse)

FOTP-15 - Altitude/Immersion of Fiber Optic Components

FOTP-16 - Salt Spray (Corrosion) Test for Fiber Optic Components

FOTP-17 Maintenance Aging of Fiber Optic Connectors and Terminated Cable Assemblies

FOTP-18 Acceleration Testing for Components and Assemblies

FOTP-20 IEC 60793-1-46 Optical Fibres - Part 1-46: Measurement Methods and Test Procedures - Monitoring of Changes in Optical Transmittance

FOTP-21 Mating Durability for Fiber Optic Interconnecting Devices

FOTP-22 Ambient Light Susceptibility of Components

FOTP-23 Air Leakage Testing for Fiber Optic Component Seals

FOTP-24 Water Peak Attenuation Measurement of Single-Mode Fibers

FOTP-25 - Repeated Impact Testing of Fiber Optic Cables and Cable Assemblies

FOTP-26 Crush Resistance of Fiber Optic Interconnecting Devices

FOTP-27 Fiber Diameter Measurements

FOTP-28 - Measuring Dynamic Strength and Fatigue Parameters of Optical Fibers by Tension

FOTP-29 Refractive Index Profile (Transverse Interference Method)

FOTP-30 - Frequency Domain Measurement of Multimode Optical Fiber Information Transmission Capacity (withdrawn May, 2003)

FOTP-31 - Proof Testing Optical Fibers by Tension (2004) (R 2005)

FOTP-32 - Fiber Optic Circuit Discontinuities

FOTP-33 Optical Fiber Cable Tensile Loading and Bending Test

FOTP-34 Interconnection Device Insertion Loss Test

FOTP-35 Fiber Optic Component Dust (Fine Sand) Test

FOTP-36 Twist Test for Connecting Devices

FOTP-37 - Low or High Temperature Bend Test for Fiber Optic Cable (ANSI/TIA/TIA- 455-37-A-93) (R2000) (R 2005)

FOTP-39 - Fiber Optic Cable Water Wicking Test (ANSI/TIA/TIA-455-39B-99) (R 2005)

FOTP-40 Fluid Immersion, Cables

FOTP-41 Compressive Loading Resistance of Fiber Optic Cables

FOTP-42 Optical Crosstalk in Components

FOTP-43 Output Near Field Radiation Pattern Measurement of Optical Waveguide Fibers

FOTP-44 Refractive Index Profile (Refracted Ray Method)

FOTP-45 Microscopic Method for Measuring Fiber Geometry of Optical Waveguide Fibers

FOTP-46 Spectral Attenuation Measurement (Long Length Graded Index Optical Fibers)

FOTP-47 Output Far Field Radiation Pattern Measurement

FOTP-48 - Measurement of Optical Fiber Cladding Diameter Using Laser-

Based Instruments (ANSI/TIA/TIA-455-48B-90) (R2000) (R 2005)

FOTP-49 Measurement for Gamma Irradiation Effects on Optical Fiber and Cables

FOTP-50 Light Launch Conditions for Long- Length Graded-Index Optical Fiber Spectral Attenuation

FOTP-50 Light Launch Conditions of Long- Length Graded-Index Optical Fiber Spectral Attenuation Measurements

FOTP-51 Pulse Distortion Measurement of Multimode Glass Optical Fiber Information Capacity

FOTP-53 Attenuation by Substitution Measurement for Multimode Graded-Index Optical Fibers of Fiber Assemblies Used in Long Length Communications Systems

FOTP-54 Mode Scrambler Requirements for Overfilled Launching Conditions to Multimode Fibers

FOTP-55 - Methods for Measuring the Coating Geometry of Optical Fibers (withdrawn July, 2000)

FOTP-56 - Test Method for Evaluating Fungus Resistance of Optical Waveguide Fiber

FOTP-56 - Test Method for Evaluating Fungus Resistance of Optical Fiber and Cable (2004) (R 2005)

FOTP-57 - Preparation and Examination of Optical Fiber Endface for Testing Purposes

FOTP-58 Core Diameter Measurements (Graded Index Fibers)

FOTP-59 Measurement of Fiber Point Defects Using an OTDR

FOTP-60 - Measurement of Fiber or Cable Length Using an OTDR (superceded by TIA- 455-133-A)

FOTP-61 - Method for Measuring the Effects of Nuclear Thermal Blast on Optical Waveguide FIber

FOTP-61 Measurement of Fiber or Cable Attenuation Using an OTDR

FOTP-62 IEC 60793-1-43 Measurement Methods and Test Procedures - Numerical Aperture

FOTP-62 IEC 60793-1-47 Measurement Methods and Test Procedures - Macrobending Loss

FOTP-63 Torsion Test for Optical Fiber

FOTP-64 Procedures for Measuring Radiation-Induced Attenuation in Optical Fibers and Optical Cables

FOTP-65 Flexure Test for Optical Fiber

FOTP-66 Test Method for Measuring Relative Abrasion Resistance

FOTP-67 - Procedure for Assessing High Temperature Exposure Effects on Optical Characteristics of Optical Fibers

FOTP-67 IEC 60793-1-51 Optical Fibres - Part 1-51: Measurement Methods and Test Procedures - Dry Heat

FOTP-68 Optical Fiber Microbend Test Procedure

FOTP-69 - Test Procedure for Evaluating the Effect of Minimum and

Maximum Exposure Temperature on the Optical Performance of Optical Fibers (ANSI/TIA/TIA-455-69A-91) (R2000)

FOTP-69 Test Procedure for Evaluating the Effect of Minimum and Maximum Exposure Temperatures on the Optical Performance of Optical Fibers

FOTP-70 - Procedure for Assessing High Temperature Exposure Effects on Mechanical Characteristics of Optical Fibers (superceded by ANSI/TIA-455-67-A)

FOTP-71 - Procedure to Measure Temperature-Shock Effects on Fiber Optic Components (ANSI/TIA/TIA-455-71-A-99)

FOTP-72 Procedure for Assessing Temperature and Humidity Cycling Exposure Effects on Optical Characteristics of Optical Fibers

FOTP-73 Procedure for Assessing Temperature and Humidity Cycling Exposure Effects on Mechanical Characteristics of Optical Fibers

FOTP-74 IEC 60793-1-53 Optical Fibres - Part 1-53: Measurement Methods and Test Procedures - Water Immersion

FOTP-75 Fluid Immersion Aging Procedure for Optical Fiber Mechanical Properties

FOTP-76 - Method of Measuring Dynamic Fatigue of Optical Fibers by Tension (withdrawn July, 2000)

FOTP-77 - Procedures to Qualify a Higher- Order Mode Filter for Measurements on Single-Mode Fiber (withdrawn May, 2003)

FOTP-78 IEC 60793-1-40 Optical Fibres - Part 1-40: Measurement Methods and Test Procedures - Attenuation

OTDR (ANSI/TIA/TIA-455-8-2000)

FOTP-80 IEC 60793-1-44 Measurement Methods and Test Procedures - Cut-off Wavelength

FOTP-81 - Compound Flow (Drip) Test for Filled Fiber Optic Cable (ANSI/TIA/TIA- 455-81B-91) (R2000)

FOTP-82 - Fluid Penetration Test for Fluid-Blocked Fiber Optic Cable (1991)

FOTP-83 Cable to Interconnecting Device Axial Compressive Loading

FOTP-84 Jacket Self-Adhesion (Blocking) Test for Cables

FOTP-85 Fiber Optic Cable Twist Test

FOTP-86 Fiber Optic Cable Jacket Shrinkage

FOT'P-87 Fiber Optic Cable Knot Test

FOTP-88 Fiber Optic Cable Bend Test

FOTP-89 Fiber Optic Cable Jacket Elongation and Tensile Strength Test

FOTP-91 Fiber Optic Cable Twist-Bend TestFOTP-88 - Fiber Optic Cable Bend Test (ANSI/TIA/TIA-455-88-2001)

FOTP-92 - Optical Fiber Cladding Diameter by Frizeau Interferometry (superceded by TIA-455-176-A)

FOTP-93 - Cladding Diameter by Non- Contacting Michelson Interferometry (withdrawn July, 2000)

FOTP-94 Fiber Optic Cable Stuffing Tubing Compression

FOTP-95 - Absolute Optical Power Test for Optical Fibers and Cables (ANSI/

TIA/TIA- 455-95-A-2000) (R 2005)

FOTP-96 Fiber Optic Cable Long-Term Storage Temperature Test for Extreme Environments

FOTP-98 Fiber Optic Cable External Freezing Test

FOTP-99 Gas Flame Test for Special Purpose Cable

FOTP- I00 Gas Leakage Test for Gas Blocked Cable

FOTP- I01 Accelerated Oxygen Test

FOTP-102 Water Pressure Cycling

FOTP-104 Fiber Optic Cable Cyclic Flexing Test

FOTP-106 - Procedure for Measuring Near- Infrared Absorbance Coating Material (withdrawn September, 2002)

FOTP-107 – Determination of Component Reflectance or Link/System Return Loss Using a Loss Test Set (2004)

FOTP-111 IEC 60793-1-34 Optical Fibres - Part 1-34: Measurement Methods and Test Procedures - Fibre Curl

FOTP-115 Spectral Attenuation Measurement of Step-Index Multimode Optical Fibers

FOTP-119 - Coating Geometry Measurement of Optical Fiber by Gray-Scale Analysis (withdrawn July, 2000)

FOTP-120 - Modeling Spectral Attenuation on Optical Fiber (superceded by TIA-455- 78-B)

FOTP-122 Polarization Mode Dispersion Measurement for Single Mode Optical Fibers by Stokes Parameter Evaluation

FOTP-123 - Measurement of Optical Fiber Ribbon Dimensions (ANSI/TIA/ TIA-455-123- 2000) (R 2005)

FOTP-124 - Polarization-Mode Dispersion Measurement for Single-Mode Optical Fibers by Interferometry Method

FOTP-124 - Polarization-Mode Dispersion Measurement for Single-Mode Optical Fibers by Interferometry.

FOTP-126 - Spectral Characterization of LEDs (ANSI/TIA/TIA-455-126-2000)

FOTP-127, Spectral Characterization of Laser Diodes

FOTP-130 - Elevated Temperature Life Test for Laser Diodes (ANSI/TIA/TIA-455-130- 2001)

FOTP-131 - Measurement of Optical Fiber Ribbon Residual Twist (ANSI/TIA/ TIA-455- 131-97)

FOTP-132 - Measurement of the Effective Area of Single-Mode Optical Fiber (ANSI/TIA/TIA-455-132-2001)

FOTP-133-A IEC 60793-1-22 Measurement Methods and Test Procedures - Length Measurement

FOTP-141 - Twist Test for Optical Fiber Ribbons (ANSI/TIA/TIA-455-141-1999) (R 2005)

FOTP-157 - Measurement of Polarization Dependent (PDL) of Single-mode Fiber Optic Components (ANSI/TIA/TIA-455-157-1995) (R2000)

FOTP-158 Measurement of Breakaway Frictional Face in Fiber Optic

Connector Alignment Sleeves

FOTP-160 IEC 60793-1-50 Optical Fibres - Part 1-50: Measurement Methods and Test Procedures - Damp Heat (Steady State)

FOTP-161 - Procedure for Assessing Temperature and Humidity Exposure Effects on Mechanical Characteristics of Optical Fibers (superceded by ANSI/TIA-455-160-A)

FOTP-162 Optical Fiber Cable Temperature-Humidity Cycling

FOTP-164 Measurement of Mode Field Diameter by Far-Field Scanning (Single-mode)

FOTP-165 - Single-Mode Fiber Diameter by Near-Field Scanning Technique (withdrawn July, 2000)

FOTP-166 Transverse Offset Method

FOTP-167 - Mode Field Diameter Measurement - Variable Aperture Method in Far-Field (withdrawn September, 2002)

FOTP-168 - Chromatic DIspersion Measurement of Multimode Graded-Index and Single-mode Optical FIbers by SPectral Group Delay Measurement in the Time Domain (superceded by TIA-175-B)

FOTP-169 - Chromatic Dispersion Measurement of SIngle-mode Optical Fibers by the Phase-shift Method (superceded by TIA-455-175-B)

FOTP-170 Cable Cutoff Wavelength of Single-mode Fiber by Transmitted Power

FOTP-171 - Attenuation by Substitution Measurement for Short-Length Multimode Graded-Index and Single-Mode Optical Fiber Cable Assemblies (ANSI/TIA/TIA-455-171-A- 2001)

FOTP-172 Flame Resistance of Firewall Connector

FOTP-173 Coating Geometry Measurement of Optical Fiber, Side-View Method

FOTP-174 Mode Field Diameter of Single-mode Fiber by Knife-Edge Scanning in Far-Field

FOTP-175 IEC 60793-1-42 Measurement Methods and Test Procedures - Chromatic Dispersion

FOTP-176-A IEC 60793-1-20 Measurement Methods and Test Procedures - Fibre Geometry

FOTP- 177 Numerical Aperture Measurement of Graded-Index Fiber

FOTP-178 IEC 60793-1-32 Optical Fibres - Part1-32: Measurement Methods and Test Procedures - Coating Strippability

FOTP- 179 Inspection of Cleaved Fiber End Faces by Interferometry

FOTP-180 - Measurement of the Optical Transfer Coefficients of a Passive Branching Device (Coupler)

FOTP-181 - Lightning Damage Susceptibility Test for Fiber Optic Cables with Metallic Components (ANSI/TIA/TIA-455-181-92) (R2001)

FOTP-183 - Hydrogen Effects on Optical Fiber Cable (ANSI/TIA/TIA-455-183-2000) (R 2005)

FOTP-184 - Coupling Proof Overload Test for Fiber Optic Interconnecting

Devices (ANSI/TIA/TIA-455-184-91) (R95) (R99)

FOTP-185 - Strength of Coupling Mechanism for Fiber Optic Interconnecting Devices (ANSI/TIA/TIA-455-185-91) (R95) (R99)

FOTP-186 - Gauge Retention Force Measurement for Fiber Optic Components (2004)

FOTP-187 - Engagement and Separation Force Measurement of Fiber Optic Connector Sets (2004)

FOTP-188 Low-Temperature Testing for Components

FOTP-189 Ozone Exposure Test for Fiber Optic Components

FOTP-190 Low Air Pressure (High Altitude) Test for Components

FOTP-191 IEC 60793-1-45 Optical Fibres - Part 1-45: Measurement Methods and Test Procedures - Mode Field Diameter

FOTP-193 - Polarization Crosstalk Method For Polarization Maintaining Optical Fiber And Components

FOTP-194 - Measurement of Fiber Pushback in Optical Connectors (ANSI/TIA/TIA-455- 194-99)

FOTP-195 IEC 60793-1-21 Optical Fibres - Part 1-21: Measurement Methods and Test Procedures - Coating Geometry

FOTP-196 - Guideline for Polarization-Mode Meaurement in Single-Mode Fiber Optic Components and Devices (ANSI/TIATIA-455- 196-99)

FOTP-197 - Differential Group Delay Measurement of Single-mode Components and Devices by the Differential Phase Shift Method (ANSI/TIA/TIA-455-197-2000)

FOTP-198 Measurement of Polarization Sensitivity of Single-Mode Fiber Optic Components by Matrix Calculation Method

FOTP-199 In-line Polarization Crosstalk Measurement Method for Polarization - Maintaining Optical Fibers Components and Systems

FOTP-200 - Insertion Loss of Connectorized Polarization-Maintaining Fiber or Polarizing Fiber Pigtailed Devices and Cable Assemblies (ANSI/TIA/TIA-455-200- 2001)

FOTP-201 - Return Loss of Commercial Polarization - Maintaining Fiber or Polarizing Fiber Pigtailed Devices and Cable Assemblies (ANSI/TIA/TIA-455-201- 2001)

FOTP-203 - Launched Power Distribution Measurement Procedure for Graded-Index Multimode Fiber Transmitters (ANSI/TIA/TIA-455-203-2001)

FOTP-204 - Measurement of Bandwidth on Multimode Fiber (ANSI/TIA/TIA-455-204- 2000)

FOTP-206 - IEC 61290-1-1 Optical Fibre Amplifiers - Basic Specification Part 1-1: Test Methods for Gain Parameters - Optical Spectrum Analyzer (ANSI/TIA/TIA-455-206- 2000)

FOTP-207 - IEC 61290-1-2 Optical Fibre Amplifiers - Basic Specification Part 102: Test Methods for Gain Parameters - Electrical Spectrum Analyzer (ANSI/TIA/TIA-455-207-2000)

FOTP-208 - IEC 61290-1-3 Optical Fibre Amplifiers - Basic Specification Part

1-3: Test Methods for Gain Parameters - Optical Power Meter (ANSI/TIA/TIA-455-208-2000)

FOTP-209 - IEC 61290-2-1 Optical Fibre Amplifiers - Basic Specification Part 2-1: Test Methods for Optical Power Parameters - Optical Spectrum Analyzer (ANSI/TIA/TIA- 455-209-2000)

FOTP-210 - IEC 61290-2-2 Optical Fibre Amplifiers - Basic Specification Part 2-2: Test Methods for Optical Power Parameters - Electrical Spectrum Analyzer (ANSI/TIA/TIA-455-210-2000)

FOTP-211 - IEC 61290-2-3 Optical Fibre Amplifiers - Basic Specification Part 2-3: Test Methods for Optical Power Parameters - Optical Power Meter (ANSI/TIA/TIA-455- 211-2000)

FOTP-212 - IEC 61290-6-1 Optical Fibre Amplifiers - Basic Specification Part 6-1: Test Methods for Pump Leakage Parameters - Optical Demultiplexer (ANSI/TIA/TIA-455- 212-2000)

FOTP-213 - IEC 61290-7-1: Optical Fibre Amplifiers - Basic Specification Part 7-1: Test Methods for Out-of-Band Insertion Losses - Filtered Optical Power Meter (ANSI/TIA/TIA-455-213-2000)

FOTP-214 - IEC 61290-1 Optical Fibre Amplifiers - Part 1: Generic Specification (ANSI TIA/TIA-455-214-2000)

FOTP-218 Measurement of Endface Geometry of Optical Connectors

FOTP-219 Multifiber Ferrule Endface Geometry Measurement

FOTP-220 - Differential Mode Delay Measurement of Multimode Fiber in the Time Domain (superceded by TIA-455-220-A)

FOTP-220 Measurement of Minimum Modal Bandwidth of Multimode Fiber Using Differential Mode Delay

FOTP-221 IEC 61290 - 5-1 Optical Fibre Amplifiers - Basic Specification - Part 5- 1: Test Method for Reflectance Parameters - Optical Spectrum Analyzer

FOTP-222 IEC 61290-3 - Optical Fibre Amplifiers - Basic Specification - Part 3: Test Methods for Noise Figure Parameters

FOTP-223 IEC 61291-2 - Optical Fibre Amplifiers - Part 2: Digital Applications - Performance Specification Template

FOTP-224 IEC 61744 Calibration of Fibre Optic Chromatic Dispersion Test Sets

FOTP-225 IEC 61745 End-Face Image Analysis Procedure for the Calibration of Optical Fibre Geometry Test Sets

FOTP-225 IEC 61745, Ed. 1.0 (1998-08): End Face Image Analysis Procedure for the Calibration of Optical Fibre Geometry Test Sets

FOTP-226 IEC 61746 Calibration of Optical Time-Domain Reflectometers (OTDR's)

FOTP-227 IEC 61300-3-24 Fibre Optic Interconnecting Devices and Passive Components - Basic Test and Measurement Procedures - Part 3-24: Examination and Measurements - Keying Accuracy of Optical Connectors for Polarization Maintaining Fibre

FOTP-228 Relative Group Delay and Chromatic Dispersion Measurement of Single-Mode Components and Devices by the Phase Shift Method
FOTP-229 - Optical Power Characterization
FOTP-231 IEC 61315 Calibration of Fibre-Optic Power Meters
FOTP-234 IEC 60793-1-52 Optical Fibres - Part 1-52: Measurement Methods and Test Procedures - Change of Temperature
FOTP-239 - Fiber Optic Splice Loss Measurement Methods
FOTP-240 Fiber Optic Connector Endface Cleaning System Evaluation

Optical Fiber System Test Procedures (OFSTPs)
(As with FOTPs, these documents are officially known as TIA-526-xxx)
OFSTP-2 - Effective Transmitter Output Power Coupled into Single-Mode Fiber Optic Cable
OFSTP-3 - Fiber Optic Terminal Equipment Receiver Sensitivity and Maximum Receiver Input
OFSTP-4 - Optical Eye Pattern Measurement Procedure
OFSTP-7 - Measurement of Optical Power Loss of Installed Single-Mode Fiber Cable Plant (2003)
OFSTP-11 Measurement Of Single-Reflection Power Penalty For Fiber Optic Terminal Equipment OFSTP-15 Jitter Tolerance Measurement
OFSTP-14 - IEC-61280-4-1(2011) (This ISO/IEC document is written for fast MM networks and may not be useful for other MM networks, so the prior version of OFSTP-14 is still considered valid), Optical Power Loss Measurement of Installed Multimode Fiber Cable Plant (1998) (r2003)
OFSTP-16 Jitter Transfer Function Measurement
OFSTP-17 Output Jitter Measurement
OFSTP-18 Systematic Jitter Generation Measurement
OFSTP-19 - Optical Signal-to-Noise Ratio Measurement Procedures for Dense Wavelength-Division Multiplexed Systems (ANSI/TIA/TIA-526-19-2000)
OFSTP-27 Procedure For System Level Temperature Cycle Endurance Test
OFSTP-28 IEC-61290-1-2: Basic Spec For Optical Fiber Amplifiers Test Methods Part 1: Test Methods For Gain Parameters - Sect. 2: Electrical Spectrum Analyzer Test Method
OFSTP-29 IEC-61290-1-3: Basic Spec For Optical Fiber Amplifiers Test Methods Part 1: Test Methods For Gain Parameters - Sect. 3: Optical Power Meter Test Method
OFSTP-30 IEC-61290-2-1: Basic Specification For Optical Fibre Amplifiers Test Methods - Part 2: Test Methods For Spectral Power Parameters - Section 2 - Optical Spectrum Analyzer Test Method

TIA Component Specifications

TIA-458-B Standard Optical Fiber Material Classes and Preferred Sizes

TIA-472 General Specification for Fiber Optic Cable

TIA-472A Sectional Specification for Fiber Optic Communication Cables for Outside Aerial Use

TIA-472B Sectional Specification for Fiber Optic Communication Cables for Underground and Buried Use

TIA-472C Sectional Specification for Fiber Optic Communication Cables for Indoor Use

TIA-472D Sectional Specification for Fiber Optic Communication Cables for Outside Telephone Plant Use

TIA-4750000-B Generic Specification for Fiber Optic Connectors

TIA-475COOO Sectional Specification for Type FSMA Connectors

TIA-475CAOO Blank Detail Specification for Optical Fiber and Cable Type FSMA, Environmental Category I TIA-475CBOO Blank Detail Specification Connector Set for Optical Fiber and Cables Type FSMA, Environmental Category 11

TIA-475CCOO Blank Detail Specification Connector Set for Optical Fiber and Cables Type FSMA, Environmental Category III

TIA-475EOOO Sectional Specification for Fiber Optic Connectors Type BFOC/2.5

TIA-475EAOO Blank Detail Specification for Connector Set for Optical Fiber and Cables, Type BFOC/2.5, Environmental Category I

TIA-475EBOO Blank Detail Specification for Connector Set for Optical Fiber and Cables, Type BFOC/2.5, Environmental Category 11

TIA-475ECOO Blank Detail Specification for Connector Set for Optical Fiber and Cables, Type BFOC/2.5, Environmental Category III

TIA-4920000-B   EN-Generic Specification for Optical Fibers

TIA-492A000-A   EN-Sectional Specification for Class Ia Graded-Index Multimode Optical Fibers

TIA-492AA00-A   EN-Blank Detail Specification for Class Ia Graded-Index Multimode Optical Fibers

TIA-492AAAA-B   EN-Detail Specification for 62.5-μm Core Diameter/125-μm Cladding Diameter Class Ia Graded-Index Multimode Optical Fibers

TIA-492AAAB-A   EN-Detail Specification for 50-μm Core Diameter/125-μm Cladding Diameter Class Ia Graded-Index Multimode Optical Fibers

TIA-492AAAC-B   EN-Detail Specification for 850-nm Laser- Optimized 50-μm Core Diameter/125-μm Cladding Diameter Class Ia Graded-Index Multimode Optical Fibers

TIA-492AAAD   EN-Detail Specification for 850-nm Laser- Optimized 50-μm Core Diameter/125-μm Cladding Diameter Class Ia Graded-Index Multimode Optical Fibers Suitable for Manufacturing OM4 Cabled Optical Fiber

TIA-492C000   EN-Sectional Specification for Class IVa Dispersion-

Unshifed Single-Mode Optical Fibers
TIA-492CA00   EN-Blank Detail Specification for Class IVa Dispersion-Unshifted Single Mode Optical Fibers
TIA-492CAAA   EN-Detail Specification for Class IVa Dispersion-Unshifted Single-Mode Optical Fibers
TIA-492CAAB   EN-Detail Specification for Class IVa Dispersion-Unshifted Single-Mode Optical Fibers with Low Water Peak
TIA-492E000   EN-Sectional Specification for Class IVd Nonzero-Dispersion Single-Mode Optical Fibers for the 1550 nm Window
TIA-492EA00   EN-Blank Detail Specification for Class IVd Nonzero-Dispersion Single-Mode Optical Fiber for the 1550 nm Window
TIA-5390000 Generic Specification for Field Portable Polishing Device for Preparation Optical Fiber
TIA-5460000 Generic Specification for a Field Portable Optical Inspection Device, Combined TIA-NECQ Specification
TIA-546A000 Sectional Specification for a Field Portable Optical Microscope for Inspection of Optical Waveguide and Related Devices
TIA-587 Fiber Optic Graphic Symbols
TIA-590 Standard for Physical Location and Protection of Below-Ground Fiber Optic Cable Plant
TIA-598 Color Coding of Fiber Optic Cables

IEC Standards
IEC Standards can be searched at the IEC website, www.iec.ch under the standards committees TC 86, SC 86A, SC 86B and SC 86C.

Standard Number
IEC 60793   Optical fibers
IEC 60794   Optical fiber cables
IEC 60869   Fiber optic attenuators
IEC 60874   Connectors
IEC 60875   Fiber optic branching devices
IEC 60876   Fiber optic spatial switches
IEC 61073   Splices for optical fibers and cables
IEC 61202   Fiber optic isolators
IEC 61274   Fiber optic adaptors
IEC 61280   Fiber optic communication subsystem basic test procedures
IEC 61281   Fiber optic communication subsystems
IEC 61282   Fiber optic communication system design guides
IEC 61290   Optical amplifier test methods
IEC 61291   Optical amplifiers
IEC 61292 TRs   Optical amplifiers technical reports
IEC 61300   Test and measurement

IEC 61313   Fiber optic passive components
IEC 61314   Fiber optic fan-outs
IEC 61751   Laser modules used for telecommunication
IEC 61753   Fiber optic interconnecting devices and passive components performance standard
IEC 61754   Fiber optic connector interfaces
IEC 61755   Fiber optic connector optical interface
IEC 61756   Fiber management system
IEC 61757   Fiber optic sensors
IEC 61977   Fiber optic filters
IEC 61978   Fiber optic passive dispersion compensators
IEC 62005   Reliability
IEC 62007   Semiconductor optoelectronic devices
IEC 62074   Fiber optic WDM devices
IEC 62077   Fiber optic circulators
IEC 62099   Fiber optic wavelength switches
IEC 62134   Fiber optic enclosures
IEC 62148   Fiber optic active components and devices – Package and interface standards
IEC 62149   Fiber optic active components and devices – Performance standards
IEC 62150   Fiber optic active components and devices – Test and measurement procedures
IEC 62343   Dynamic modules

Note: This list was assembled from a number of sources with various dates - it is not complete because they change all the time. A full catalog of TIA specs is at http://www.tiaonline.org/

FOA Standards
FOA's Standards are concise standards created by FOA with the participation of experts in the field for the most common issues affecting fiber optic network owners, contractors, designers and installers. Each standard summarizes the basic information the reader needs to know in just 1 page. Each of the FOA's Standards will reference other industry standards that are similar in scope and which are used as the basis of the FOA standard, allowing FOA standards to be substituted for them. These FOA standards can be used for reference in project paperwork when the user and contractor need to be certain they agree what is being specified for the project.

FOA Standard FOA-1: Testing Loss of Installed Fiber Optic Cable Plant, (Insertion Loss, TIA OFSTP-14, OFSTP-7, ISO/IEC 61280, ISO/IEC 14763, etc.) More

FOA Standard FOA-2: Testing Loss of Fiber Optic Cables, Single Ended, (Insertion Loss, TIA FOTP-171, OFSTP-7, , ISO/IEC 14763) More
FOA Standard FOA-3: Measuring Optical Power (Transmitter and Receiver Power, FOTP-95, Numerous ISO/IEC standards) More
FOA Standard FOA-4: OTDR Testing of Fiber Optic Cable Plant (TIA FOTP-8/59/60/61/78, ISO/IEC 14763, etc.) More
FOA Standard FOA-7: Mode Conditioning For Testing Multimode Fiber Optic Cables (Mandrel wrap, encircled flux)
All FOA standards are included at the end of this section.

ANSI/NECA/FOA-301 Installing And Testing Fiber Optics is a standard for installing and testing fiber optic cable plants. Download a free copy from the FOA website.

FOA Standard FOA-1
Testing Loss Of Installed Fiber Optic Cable Plant

This test will measure the loss of an installed fiber optic cable plant, singlemode or multimode, including the loss of all fiber, splices and connectors.

Equipment Needed To Perform This Test
1. Test source appropriate for the fiber being tested (Multimode: 850 and/or 1300nm LED, singlemode, 1310 and/or 1550 nm laser)
2. Optical power meter calibrated at the same wavelengths as the source output.
3. Launch and receive reference cables of the same fiber type and size as the cable plant and have connectors compatible to those on the cable plant. They should be tested per FOA-2 to ensure they are in good condition.
4. Mating adapters compatible to connectors
5. Cleaning supplies

Test Procedure
1. Turn on equipment and allow time to warm-up
2. Attach launch cable to source. This should remain connected to source for the duration of the test.
3. Clean all connectors and mating adapters.
4. Set "0 dB" reference using method shown to the right. Meter may be set to read "0 dB."
5. Attach source/ref cable and meter/ref cable to the cable plant under test and make loss measurement.

Options For "0 dB Reference" - Set Before Testing
1. Use the "1 Cable Reference" if connectors are the same on the cable plant as the testers and reference cables or may be adapted using hybrid adapters.
2. Use the "2 Cable Reference" if connectors are not the same on the cable plant as the testers but can be mated to each other with adapters and hybrid reference cables are being used.
3. Use the "3 Cable Reference" if connectors are plug/jack styles and reference cables are both ended in either plugs or jacks.

Reducing Measurement Uncertainty
1. Clean all connectors regularly before and while testing.
2. Use modal control on launch cable, e.g. small loop on SM fiber or mandrel wrap on MM fiber.
3. Check "0 dB Reference" periodically during testing.
4. 4. Periodically check reference cables per FOA-2 to verify their condition.

Documentation
Record the date of the test, operator, test equipment used, reference method, cable and fiber identification, test wavelength and measured loss.

Test Diagram

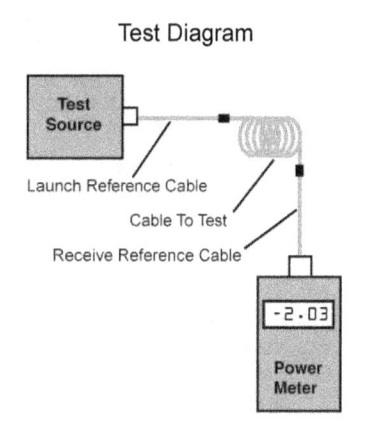

Setting "0 dB" Reference

1 Cable Reference

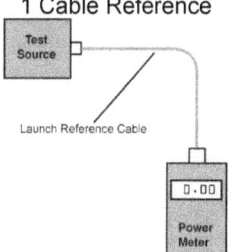

2 Cable Reference

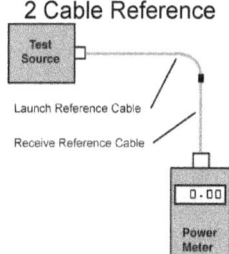

3 Cable Reference

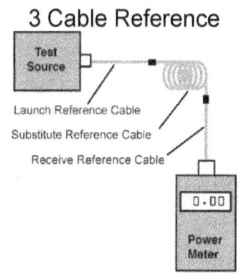

FOA Standard FOA-2
Testing Loss Of Fiber Optic Cables, Single-Ended

This test will measure the loss of a fiber optic cable, singlemode or multimode, including connectors on each end individually. For short cables, e.g. patchcords, with negligible fiber loss, the measured loss may be considered the loss of the connector mated to the reference connector.	Test Diagram 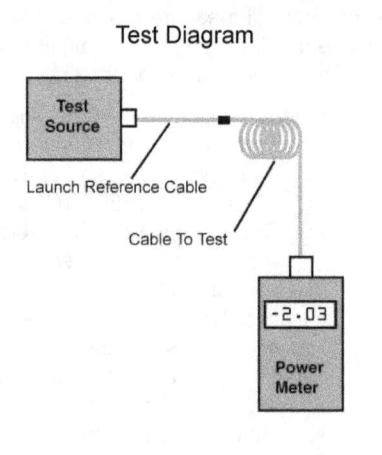

Equipment Needed To Perform This Test
  6. Test source appropriate for the fiber being tested (Multimode: 850 and/or 1300nm LED, singlemode, 1310 and/or 1550 nm laser)
  7. Optical power meter calibrated at the same wavelengths as the source output with adapters to mate to connector type on cable.
  8. Launch reference cable that is the same fiber type and size as the cable plant and have connectors compatible to those on the cable.
  9. Mating adapters compatible to connectors
  10. Cleaning supplies	

Test Procedure
  6. Turn on equipment and allow time to warm-up
  7. Attach launch cable to source. This should remain connected to source for the duration of the test.
  8. Clean all connectors and mating adapters.
  9. Set "0 dB" reference using method shown to the right. Meter may be set to read "0 dB."
  10. Attach source/ref cable and to the cable under test and make loss measurement.
  11. Reverse cable and test again.
	Setting "0 dB" Reference
	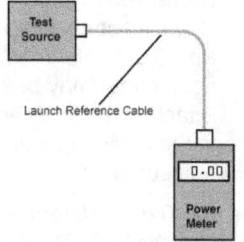

Options For Testing With Different Connector Types 4. If the connector(s) on the cables to test are "plug and jack" type and/or are not compatible to the optical power meter for testing and/or reference, you cannot test single-ended. Use the FOA-1 method with a 2 or 3 Cable Reference as appropriate. Results will include loss of connectors on both ends.	

Reducing Measurement Uncertainty
  5. Clean all connectors regularly before and while testing.
  6. Use modal control on launch cable, e.g. small loop on singlemode fiber or mandrel wrap on multimode fiber.
  7. Check "0 dB Reference" periodically during testing.	

Documentation Record the date of the test, operator, test equipment used, reference method, cable and fiber identification, test wavelength and measured loss.	

FOA Standard FOA-3
Measuring Optical Power In Fiber Optic Systems

This test will measure the optical power exiting the end of a fiber optic cable. This test is commonly used to measure the coupled power of a fiber optic source in a transmitter, power into a receiver or for setting references for optical loss measurements.	Test Diagram 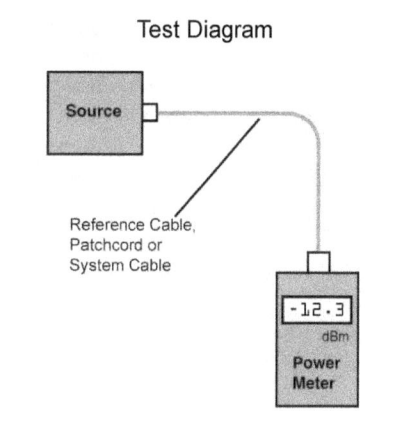 Reference Cable, Patchcord or System Cable

Equipment Needed To Perform This Test
   11. Fiber optic power meter calibrated at the same wavelength as the source output (e.g. multimode: 850 and/or 1300nm, singlemode, 1310, 1490 and/or 1550 nm, POF 650 nm) capable of measuring optical power in the power range of the source.
   12. Optical power meter adapters to mate to connector type on cable.
   13. Reference cable that is the same fiber type and size as the cable plant and have connectors compatible to those on the source and cables.
   14. Cleaning supplies	

Test Procedure
   12. Turn on meter and allow time to warm-up
   13. Set meter to wavelength of source and "dBm" to measure calibrated optical power.
   14. Clean all connectors and mating adapters.
   15. Attach reference cable to source if testing source power or disconnect cable from receiver.
   16. Attach power meter to end of cable and read measured power.
	Note:
	A reference cable or known good patchcord is used for testing source power coupled into a fiber. Receiver power is tested by disconnecting the system cable connecting to the receiver and attaching it to the power meter to measure power.

Options For Testing Power is generally measured in "dBm" or dB referenced to 1 milliwatt of optical power. Optical power measurements may also be made in Milliwatts (mW) or microwatts (µW)	

Reducing Measurement Uncertainty
   8. Calibrate optical power meter according to manufacturer specified intervals.
   9. Clean all connectors and remove meter adapter periodically to clean the adapter and power meter detector.
   10. Do not bend fiber optic cables tightly to cause stress loss.	

Documentation Record the date of the test, operator, test equipment used, cable and fiber identification, test wavelength and measured power.	

FOA Standard FOA-4
OTDR Testing of Fiber Optic Cable Plants

OTDR testing creates a snapshot of a fiber optic cable. This test is commonly used to verify the quality of the installation and troubleshoot problems. OTDR testing requires interpretation of the data acquired, called the trace or signature, by a skilled operator.

Test Diagram

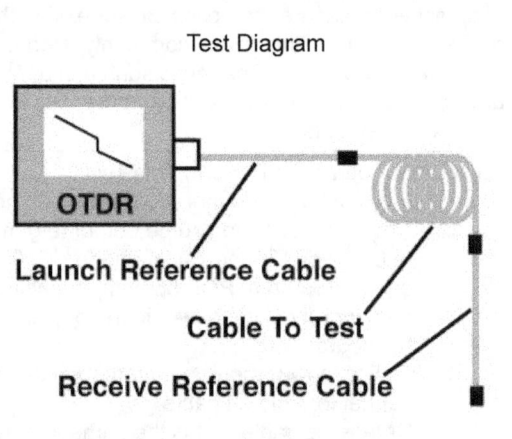

Launch Reference Cable

Cable To Test

Receive Reference Cable

Equipment Needed To Perform This Test
15. OTDR with modules appropriate for the cable plant (e.g. multimode: 850 and/or 1300nm, singlemode, 1310, 1550 and/or 1625nm.)
16. Launch and/or receive reference cables of the same fiber type and size as the cable plant and with connectors compatible to those on the cable plant.
17. Cleaning supplies

Test Procedure
17. Turn on OTDR and allow time to warm-up
18. Set parameters on OTDR appropriate for the cable plant being tested (range, wavelength, number of averages, etc.)
19. Clean all connectors and mating adapters.
20. Attach launch reference cable to OTDR and to cable plant under test.
21. Attach optional receive cable to far end of cable under test.
22. Acquire trace and analyze.

Information In OTDR Traces

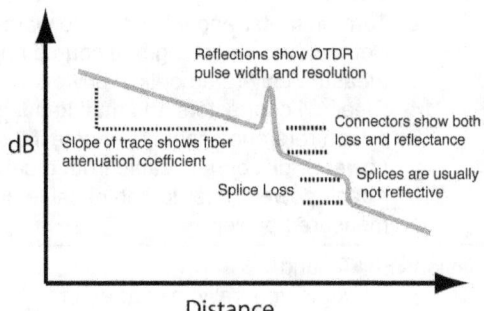

Distance

Options For Testing
1. Use of the receive reference cable is optional, it is required if the far end connector loss is to be measured and included in total cable plant loss
2. Testing at more than one wavelength may be required. Longer wavelength testing is often used to find stress related to installation problems. Traces may be compared for analysis.

Notes
1. Insertion loss testing of the cable plant is recommended for acceptance testing.
2. Not all cable plants are long enough for OTDR testing. Ensure the OTDR has sufficient resolution for the cables being tested.
3. Always use a launch cable long enough to allow the OTDR to recover from test pulse overload and permit proper testing of the cable plant.
4. Do not use the OTDR automatic cable analysis until a skilled technician analyzes a trace and confirms it is appropriate for the cable plant under test.

Documentation
Record the date of the test, operator, test equipment used, cable and fiber identification, test wavelength(s) and all traces for the fiber under test.

FOA Standard FOA-7
Mode Conditioning For Testing Multimode Cables

This standard covers mode conditioning multimode fiber optic cables for insertion loss testing per most standards. This mode conditioning will result in more consistent test conditions which will provide more accurate test results. For 50/125 fibers it will allow meeting Encircled Flux (EF) standards for mode conditioning.

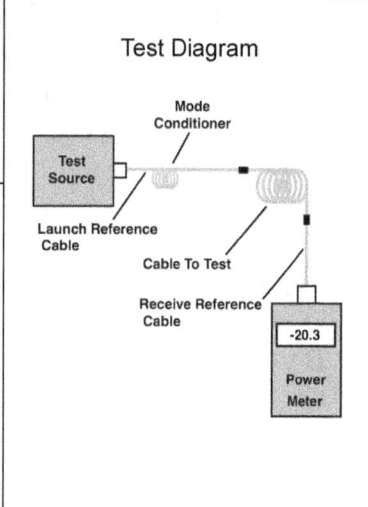

Test Diagram

Equipment Needed To Perform This Test
18. Test source appropriate for the fiber being tested: Multimode, 850 nm LED,
19. Optical power meter calibrated at the same wavelengths as the source output.
20. Launch and receive reference cables of regular MM fiber the same size as the cable plant and have connectors compatible to those on the cable plant. *Bend insensitive fibers will not work with a mandrel wrap.*
21. Mating adapters compatible to connectors
22. Mandrel of size specified below.
23. Cleaning supplies

Specified Mandrel Size – Wrap 5 Turns

Fiber/Cable Type	3mm Jacket	2 or 2.4 mm Jacket	1.6 mm Jacket	900 micron buffered fiber
50/125 micron	22 mm	23 mm	24 mm	25 mm
62.5/125 micron	17 mm	18 mm	19 mm	20 mm

Procedure
23. Wrap 5 turns of the launch cable around the mandrel at about 1/3 the distance from the test source to the meter.
24. Secure the cable with tape if necessary.

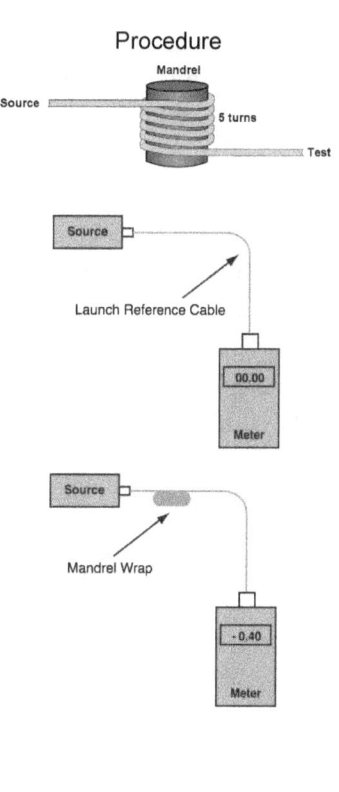

Procedure

HOML (Higher Order Mode Loss) Test For Mandrel Wrap. Testing 50/125 Fiber Cable For Encircled Flux Compliance.
5. Before installing the mandrel wrap, measure the output power of the launch cable with the power meter and record or "zero" this power.
6. Install the mandrel wrap. Tape the cable to the mandrel if necessary to secure it.
7. Measure the power after installation of the mandrel wrap. Note the difference in power before and after the mandrel wrap has been done. This is the higher order mode loss (HOML)
8. If the HOML id 0.20 to 0.60 dB, the source is EF compliant and ready to use without the mandrel. Remove the mandrel and use the source for tests.
9. If the HOML is >0.60 dB, leave the mandrel on the reference launch cable and make measurements.
10. If the HOML is <0.20 dB, the source has too low a mode fill and should not be used.

Documentation
When testing, record the date of the test, operator, test equipment used, mandrel size, reference method, cable and fiber ID, test wavelength and measured loss.

Chapter Exercises

- Go to the TIA, ISO, IEC and ITU websites and look at the types of standards written for fiber optics.
- Go to manufacturers' websites and see how they use standards to describe and promote their products.

Chapter Quiz

1. Cabling standards are generally written _by_ _____.
   - A. Manufacturers
   - B. Contractors
   - C. Installers
   - D. Users

2. Cabling standards are generally written _for_ _____.
   - A. Manufacturers
   - B. Contractors
   - C. Installers
   - D. Users

3. The best and most current source of information on standards is generally _____.
   - A. Component manufacturers
   - B. Reading the standards themselves
   - C. TIA
   - D. ISO

5. Testing a cable plant and "certifying" it per the standards means the cable plant_____.
   - A. Is tested under network operating conditions
   - B. Meets the minimal specifications required by the standards
   - C. Exceeds the performance requirements of the standards by a large margin
   - D. Tests better than the loss budget calculated for it during the design stage

Chapter 4

Fiber Optic Test Instruments

Objectives: From this chapter you should learn:
What types of instruments are required for fiber optic tests
How those instruments work
What are the specifications for these instruments
How to choose the instrument you need

Instruments Required For Fiber Optic Tests

Fiber optic testing involves many different tests using different types of test equipment. Before any actual testing begins, the technician must purchase test equipment and learn how to use them. This chapter will cover the various types of fiber optic test equipment, relevant specifications and how they are used. Following chapters will cover each category of test and how to use these instruments properly in performing the tests.

The most common measurement parameters and the required test equipment are listed in the Table below.

Fiber Optic Tests And Instrument Requirements

Test Parameter	Instrument
Visual inspection of connectors	Fiber optic inspection microscope
Visual tracing of multimode fibers	Fiber tracer
Visual tracing of singlemode fibers and visual fault location	Visual fault locator
Optical power (source output, receiver signal level)	Fiber optic power meter
Attenuation or Loss of Fibers, Cables & Connectors (Insertion Loss)	Fiber Optic Power Meter & Test Source, OLTS (optical loss test set)
Backscatter (loss, length, reflectance and fault Location)	Optical Time Domain Reflectometer (OTDR)
Fault location	OTDR, Visual cable fault locator

Bandwidth / dispersion (MM: modal & chromatic, SM: chromatic and polarization mode)	Dedicated bandwidth testers
Reflectance	OTDR, OCWR (Optical Continuous Wave Reflectometer)
Fiber geometry (Core and cladding diameter, concentricity, etc.)	Various mechanical and optical inspection tools
Source wavelength, spectral width	Spectrum analyzer

With all different test equipment and their applications, it is important to first understand the tests involved and then to choose appropriate test equipment. Once equipment is chosen the operators should get trained on using that instrument so they know how to use the instrument properly to perform the tests correctly. Then they need to learn how to interpret the data it provides in the context of those tests. That's the focus of this book.

Visual Inspection Of Connectors with Microscopes

There are two major uses for visual inspection of fiber optic connectors. Polished connector ferrules require visual inspection during manufacturing to evaluate polishing and find possible defects during the connector termination process. In the field, connectors need inspection for cleanliness and damage before testing or connection to another connector or transceiver. See Chapter 4 on cleaning and visual inspection.

Visual inspection is accomplished using a microscope that has a fixture to hold the fiber or connector steady in the field of view and a light source to illuminate the connector. Fiber optic inspection microscopes vary in magnification from 30 to 800 power, with 100-400 power being the most widely used range for connector ferrule inspection. Lower magnification is used to view more of the connector when looking for dirt or contamination. Higher magnification allows more critical evaluation of the ferrule for polish, scratches or cracks in the fiber. Lighting is usually provided two ways, direct along the axis of the connector ferrule and at an angle to the ferrule end to more clearly show dirt and scratches.

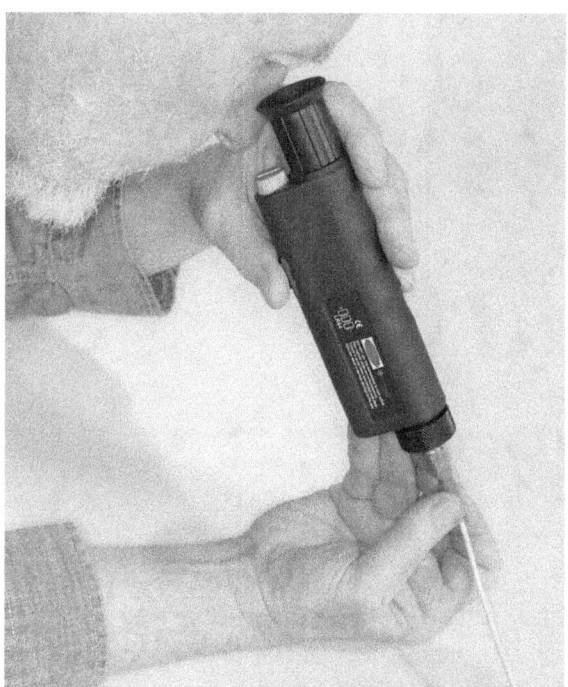

Portable optical microscope used for connector inspection

Optical image of connector ferrule end showing dirt and scratches

Microscopes can be optical with observation by the eye or video with an image displayed on a small screen on the instrument, a PC, smartphone or tablet computer. Video microscopes cost more but they generally have software that can help evaluate the condition of the end face of the connector ferrule according to international standards. Video microscopes also allow for storing an image of the connector, valuable for documenting the condition of the connector at installation and for reference in the future.

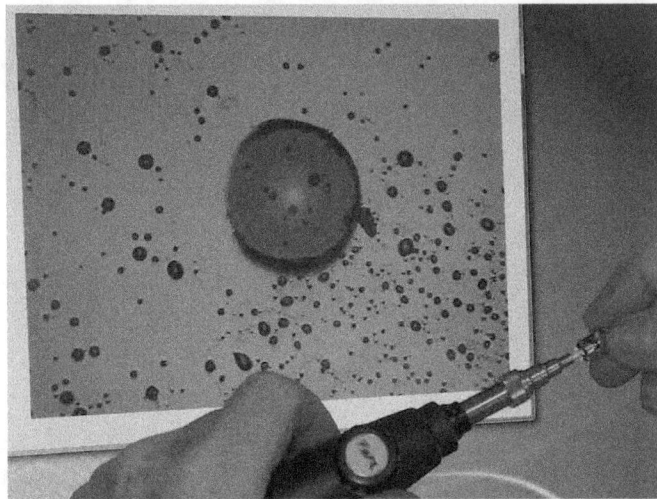
Video microscope view of dirty connector

Visual microscopes should have a built-in infrared filter to remove any signals being transmitted in the fiber for the protection of the eye of the user. An optical microscope can magnify any light in the fiber and focus it in the eye, a potential danger, so an infrared filter is used for protection. A video microscope is preferable since it has no danger to the eye.

Fiber Optic Cleaning Kits

While technically not "instruments," fiber optic cleaning kits are an essential part of every fiber optic tech's tool and testing equipment inventory. Dirt and contamination is the most common problem encountered in fiber optics, either with testing or with connecting communications systems. Dirt and contamination causes loss when making connections and may contaminate or damage mating connectors or active device ports.

Cleaning began with using lab grade isopropyl alcohol and lint-free wipes but has evolved into a much more scientific process with specialized cleaners of many types. The importance and complexity of cleaning is the reason that a complete chapter will be dedicated to the topic in this book.

Cleaning kits are available that offer comprehensive cleaning products for both wet and dry cleaning for most types of connectors

Inspection for cleanliness has also become more sophisticated. Besides inspecting the end of the ferrule on a connector, tools are now available to look at the entire mating end of the connector for dirt that may migrate into the connection and cause problems. All this will be covered in the next chapter.

Fiber optic cleaning kit

Visual Cable Tracers and Fault Locators

Many of the problems in connection of fiber optic networks are related to making proper connections, usually referred to as polarity. Since the light used in systems is infrared and invisible, one cannot see the system transmitter light, and of course, for safety one should never look into a fiber without testing for the presence of an optical signal with a power meter first.

By injecting the light from a visible source, such as a LED or incandescent bulb for multimode fiber or a visible laser for singlemode fiber, one can visually check the continuity of a fiber and trace the fiber from transmitter to receiver to insure correct polarity. The simple instruments that inject visible light are called fiber tracers and visual fault locators (VFL).

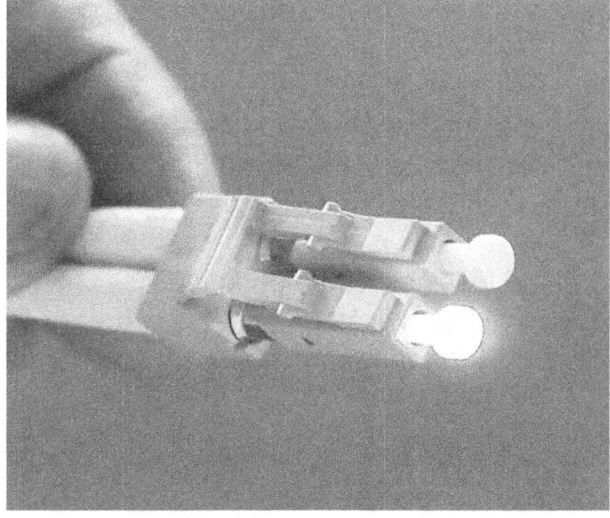

Visual tracer or fault locator identifies fibers

If a powerful enough visible light, such as a visible diode laser is injected into the fiber, you can trace fibers for 2 to 5 kilometers. On simplex fiber patchcords you can also see high loss points caused by stress in the cable. Most applications center around short cables such as patchcords used in LANs and telco central offices to connect equipment to the fiber optic cables.

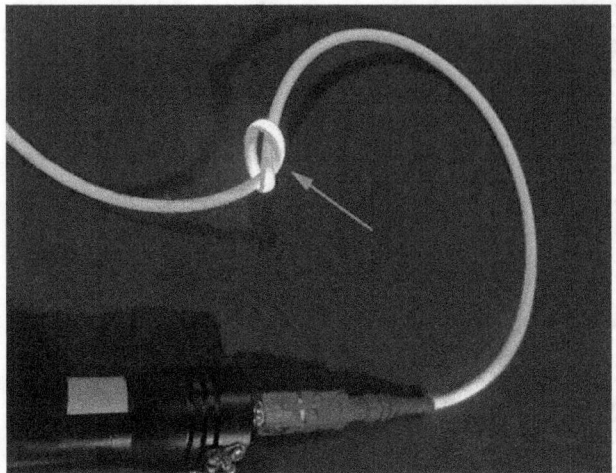

Visual fault locator shows stress loss in cable

The VFL works best nearer the instrument which covers the range where other instruments like optical time domain reflectometers (OTDRs) are not useful, so it is complementary to the OTDR in cable troubleshooting (see OTDRs below).

This method will work on buffered fiber and even jacketed single fiber cable if the jacket is not opaque to the visible light. The yellow jacket of singlemode fiber and orange of multimode fiber will usually pass the visible light. Most other colors, especially black and gray, will not work with this technique, nor will most multifiber cables.

VFLs are particularly useful in identifying fibers in splice closures and finding bad splices in splice trays. In subdued lighting, the visible laser light will make it easy to identify fibers or problem splices.

VFLs operate at around 650nm, not in the low loss range of optical fiber. Since the loss in the fiber is quite high at visible wavelengths, on the order of 9-15 dB/km, this instrument has a limited useful range, typically 3-5 km.

Fiber Optic Power Meters

Optical power is the most important test in fiber optics, similar to voltage in electrical measurements. Power in fiber optics is measured by a fiber optic power meter, also called an optical power meter. Optical power tests are used to evaluate transceiver source power output and receiver power input for system tests. A power meter is used with a test source to measure the loss of the cable plant. Power measurements are used in all other instruments that measure loss, reflectance or dispersion.

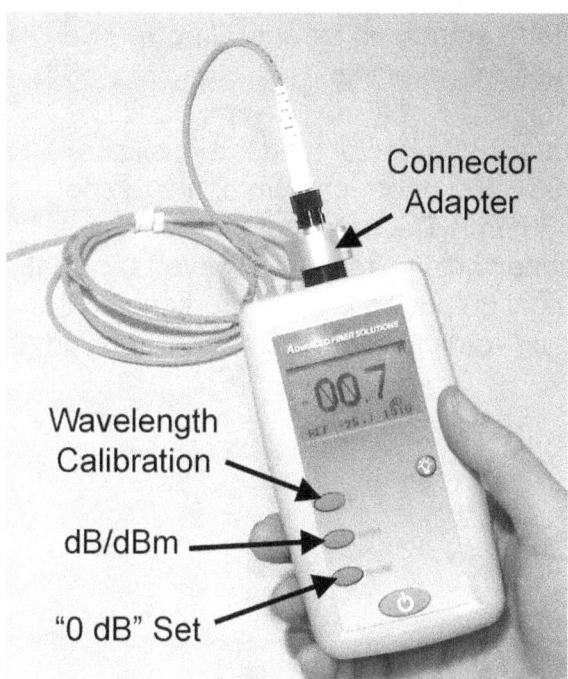

A typical fiber optic power meter showing controls

The instrument designed to measure power is called a fiber optic power meter. It is essentially a special light meter with digital display calibrated at the wavelengths used in fiber optics, typically 850, 1300 and 1550 nanometers (nm), and sometimes 650 nm for plastic optical fibers. To interface to the variety of fiber optic connectors in use, some form of removable connector adapter is usually provided.

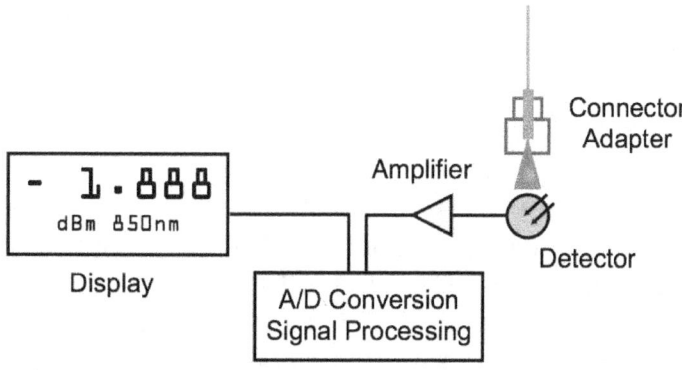

Block diagram of fiber optic power meter

All power meters measure power in dBm (decibels referenced to 1 milliwatt) and most also measure in dB (relative to an arbitrary power level set by the user, usually to measure loss.) Some meters also have a linear scale in Watts, with ranges scaled in milliwatts to correspond to the dBm range. The meter is calibrated to international transfer standards for optical power. See Chapter 4 for details on the calibration of optical power.

Power meters typically have a solid state light detector made from silicon for short wavelength systems (650 to 850 nm) or germanium or InGaAs for longer wavelength systems (850 to 1550 nm.) The output of the detector is connected to an amplifier and signal conditioning circuitry to convert the light input to an electrical output. The electrical output which is proportional to the light power input is digitized by an analog-to-digital converter and displayed on a digital display.

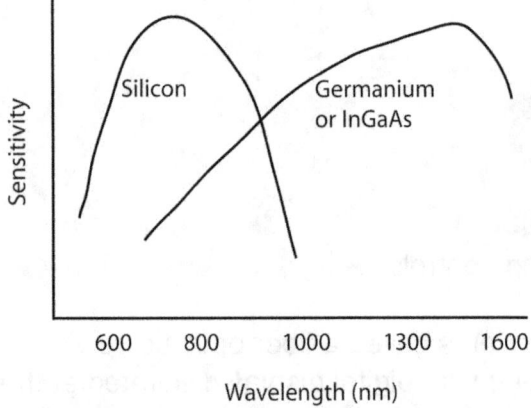

Sensitivity of power meter detectors

Solid state detectors do not have the same sensitivity to all wavelengths of light so the meter must be calibrated at each wavelength separately. See the section on calibration in Chapter 4. The difference in sensitivity as shown in the graph above is significant. When measuring power, it is important to set the meter to the correct wavelength on its control panel.

Power meter detectors need to be large enough to capture all the light exiting the optical fiber being measured to ensure a proper measurement. The cone of light can vary considerably between OM1 fiber with a large core diameter and high numerical aperture and a singlemode fiber. The larger the detector the more likely it will capture all the light, but there are tradeoffs. Larger detectors have more noise and therefore cannot measure lower light levels. And larger detectors are more expensive, making them less likely to be used in low cost meters.

Power meters display power in decibels (dB.) One scale is dB referenced to one milliwatt of optical power (dBm) that is used for *absolute* power measurements such as the output of an optical transmitter or the input to a receiver. Meters have a relative dB scale also, used for loss measurements that are a *relative* measurement of power. When measuring loss, the reference value may be set to "0 dB" at any power level on the output of the test source for convenience. Occasionally, lab meters may also measure in linear units (milliwatts, microwatts and nanowatts.) The relationship of power in dB to milliwatts is logarithmic, based on $\log_{10}$. DB will be explained in detail in the next chapter.

Power meters cover a very broad dynamic range of power, typically over 1 million to 1 or 60 dB. Although most fiber optic power and loss measurements are made in the range of +10 dBm to -50 dBm, some power meters offer much wider dynamic ranges, especially if they are designed to measure the output of high power laser sources and fiber amplifiers.

For testing analog CATV systems or fiber amplifiers, one needs special meters with extended high power ranges up to +20 dBm (100 mW). Although no fiber optic systems operate at very low power, below about -50 dBm, some lab meters offer ranges to -70 dBm or more, which can be useful in measuring optical return loss or spectral loss characteristics with a monochromator source.

Power meters measure the time average of the optical power, not the peak power, so the meters are sensitive to the duty cycle of a modulated input. A digital pulse signal, for example, may vary somewhat in how many "1s" and "0s" are being transmitted, but over time the average remains fairly constant.

When measuring the output of a transceiver or network equipment, the manufacturer will generally specify power levels for typical operating conditions so the tech testing it does not have to make any corrections to the meter reading.

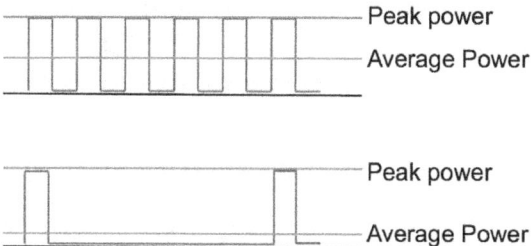

Average power is a function of duty cycle of the signal

If necessary, one can calculate peak power if one knows the duty cycle of the input, by dividing the average power by the duty cycle. For loss measurements, one uses a fiber optic test source with CW (steady state) or 2 kHz pulsed output. As long as the source modulation doesn't change, no compensation needs to be made for duty cycle.

Specifications For Fiber Optic Power Meters
Manufacturers provide basic data on a power meter to allow evaluating the product for your application. Typical specifications will include at least the following items.

Optical Input/Measurement Ranges
Detector type: Silicon, Germanium or InGaAs
Calibration wavelengths (nm): 850, 1300, 1550 (optional, 650, 1490, etc.)
Power ranges: +10 to -70 dBm, 0.01μw to 10mw (optional)

Fiber Optic Connector Interface:
Modular adapter system compatible with ST, SC, LC connectors (optional, others)

Performance
Accuracy: +/-5%(W), +/-0.2 dB
Temperature coefficient: 0.1%/°C
Operating temperature range: -10 to 50 °C
Storage temperature range: -30 to 60 °C
Battery life (hrs), recharge time

Display:
Liquid crystal display
Selectable resolution: 0.1 or 0.01 dB
Shows power reading, range, calibration wavelength and low battery

Controls:
Power: ON/OFF
Range: dB/dBm/mW
Calibration wavelength

Important specifications to consider when purchasing a fiber optic power meter include wavelength calibrations, power range, accuracy, temperature ranges and battery life. You may also want to choose a meter with data storage that can be downloaded to create reports without manual recording of data.

Fiber Optic Test Sources

In order to make direct measurements of optical loss or attenuation in fibers, cables and connectors, what is called insertion loss, one must have a fiber optic test source as well as a power meter. The fiber optic test source, also called a light source is intended to simulate the transmitter source used in actual systems. It must be chosen for compatibility with the type of fiber being tested (singlemode or multimode with the proper core diameter) and the wavelength desired for performing the test.

Most test sources are either LED's or lasers of the types commonly used as transmitters in actual fiber optic systems, making them representative of actual applications and enhancing the usefulness of the testing. An exception to this is the sources used with multimode fiber are generally specified in standards to be LEDs while actual systems now primarily used a type of laser, a VCSEL or vertical cavity surface emitting laser. VCSELs are rarely used in test sources and never specified in standards because of concerns that they are not predictable in modal distribution when coupled to fiber, a major factor in measuring loss in multimode fiber. Modal distribution in test sources is discussed below.

Typical wavelengths of sources for multimode fiber are 650 nm for plastic fiber, 850 nm for glass fiber at short wavelengths and 1300 nm for long wavelengths. Multimode sources will generally have LEDs at 850 nm only or 850 and 1310nm for dual wavelength testing. Few multimode systems operate at 1300 nm today; most actually use 850 nm VCSELs or vertical cavity surface-emitting lasers. Testing may still be required by older standards but is mostly unnecessary.

Sources for singlemode fiber are typically specified at 1310 and 1550 nm. The discrepancy in wavelengths at 1300 for LEDs and 1310 nm for lasers is not technical but historical. Laser and LED wavelengths are a matter of the composition of the semiconductor chips that make the lasers. The lasers at this wavelength were centered at 1310 nm and had very narrow spectral widths (the variety of wavelengths of the light emitted by the laser) so an exact wavelength was used. LEDs have very wide spectral widths, so the wavelength was rounded off at 1300 nm. The nomenclature has been in use ever since.

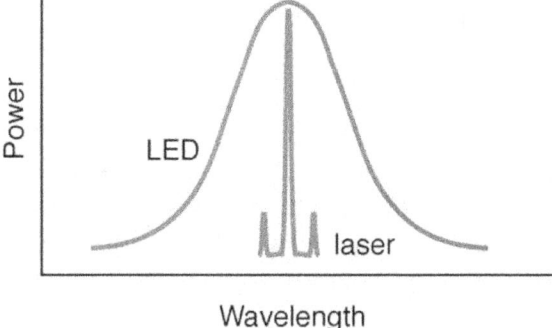

The source wavelength can be a critical issue in making accurate loss measurements on long links of singlemode fiber and even short links of multimode fiber, since the attenuation of the fiber is highly sensitive especially to wavelength at short wavelengths. Thus all test sources should be near the nominal wavelength or calibrated for wavelength.

Another source-related factor affecting measurement accuracy is the stability of the output power. To make accurate measurements the source output must be stable over typical changes in temperature and condition of the power supply, especially those that are battery powered.

One problem in some portable sources is they vary in optical power when the batteries discharge. Sources should be designed with power stabilization to prevent this instability.

For extremely accurate measurements, the source may use optical feedback stabilization to maintain output power at a precise level for long times required for some measurements. Most lasers are optically stabilized. Under any circumstances, sources should be turned on and allowed to warm up for several minutes to stabilize before making measurements.

When using a source to test loss of a fiber optic cable plant, patchcord or other components, the source will be coupled to a reference launch cable which allows it to connect to the cable under test. A source will generally have a fixed connector port for the source output so a hybrid reference cable will be needed when testing cables with incompatible connectors. For multimode fiber, modal distribution in the test source and its attached launch cable can affect measurements. This topic will be discussed in Chapter 6 on insertion loss.

Industry standards have requirements or recommendations on the modal output of test sources for multimode fiber. Few multimode LED test sources are made with specified modal distribution so external devices called mode scramblers, filters and strippers may be used with a launch cable to adjust the modal distribution in the fiber to approximate actual operating conditions. Modal control will be discussed in detail in Chapter 6 on Insertion Loss.

Test sources almost always have fixed connectors. Hybrid test jumpers with connectors compatible with the source on one end and the connector of the cable being tested on the other must be used as reference cables. This may affect the type of reference setting mode used for loss testing. Chapter 6 will cover this in more detail.

Some laboratory tests, such as measuring spectral attenuation of fiber, require a variable wavelength source, which is usually a tungsten lamp with a monochromator to vary the output wavelength. These tests are generally limited to manufacturers of fiber and use special test equipment made to characterize fibers as manufactured.

Specification For Fiber Optic Test Sources
Manufacturers provide basic data on a test source to allow evaluating the product for your application. Typical specifications will include at least the following items. Actual specifications are for illustration only

Optical Source Output:
Source Type (LED, laser, VCSEL
Wavelength (850, 1300 nm LEDs, 850 nm VCSEL, 1310, 1550 nm laser
Wavelength Tolerance: 850 nm ±20 nm
Spectral Width (LED)

Controls:
On/Off : power switch
Wavelength (multiple wavelength sources)
Adjust output power level (optional)
Modulation: CW/2 kHz modulation on lasers

Environmental:
Operating temperature range: -10 to 40 deg C
Storage temperature range: -40 to 70 deg C
Temperature coefficient of drift (LED optical power): 0.5%/deg C
Stability (laser, optical feedback stabilized): 0.02 dB/hr

Power:
AC line or rechargeable battery.
Battery life 8 hours at full optical power output,
Recharge time: 16 hours to full recharge

Optical Loss Test Sets/Test Kits

The optical loss test set (OLTS) is an instrument formed by the combination of a fiber optic power meter and fiber optic test source that is primarily used to measure the loss of fiber optic cable plants. Early versions of this instrument were called attenuation meters. In some standards, an OLTS is called LSPM for light source and power meter. What some manufacturers call a test kit has a similar purpose, but is usually comprised of separate instruments and includes accessories to customize it for a specific application, such as testing a FO LAN, telco or CATV. Some UTP copper testers offer modules that convert the copper tester into an OLTS.

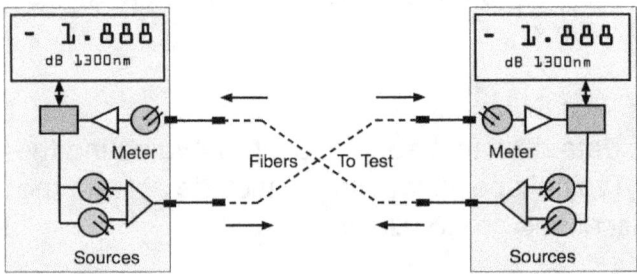

Block diagram of a typical optical loss test set

The OLTS will have a fiber optic power meter and display plus one or more sources, typically two. The sources are generally connected to a fiber optic coupler that connects the two sources to one output. The OLTS has a controller that switches on one of the sources, calibrates the meter to the correct wavelength, stores the 0 dB reference when the instrument is calibrated, then measures, displays and records the results of a test.

Calibrating the 0 dB reference is done with two instruments brought together or by looping a single reference jumper from the source output to the meter input.

When making a measurement, the instruments will be connected to the fibers to test with reference test cables and a test begun. It will test both wavelengths on both fibers and store date for downloading on a computer.

The OLTS may have several optional features that affect its use. Some have separate source outputs and meter inputs like a separate power meter and test source as shown above, with two wavelengths from one source output (MM: 850 and 1300 nm, SM: 1310 and 1550 nm.) Some OLTSs combine the meter and source ports into one port to offer bidirectional testing on a single fiber and some have two bidirectional ports to allow testing two fibers at once. Bidirectional port may also have problems meeting standards for modal power distribution in multimode fibers. Ask the OLTS manufacturer about these issues.

An OLTS will generally have fixed connector ports for both source output and meter input so hybrid reference cables will be needed when testing cables with incompatible connectors.

Some manufacturers of premises copper cabling testers offer modules to convert these testers to an OLTS, allowing fiber and copper testing with one instrument.

Compared to a separate test source and power meter, the combination OLTS instrument may be more expensive than an individual source and power meter but offer advantages in testing time. Since the ends of an installed fiber cable are usually separated by some distance, testing requires two OLTS units instead of one source and one meter. Some OLTS may offer different instruments for each end of the cable that operate in a master and slave mode.

Specifications for an OLTS will be a combination of the specifications for a power meter and a test source, plus it should cover controls and interfaces that affect operation.

Reference Test Jumper Cables and Mating Adapters

In order to test cables in an insertion loss test, one needs to establish test conditions. This requires a reference launch jumper cable to connect the test source to the cable under test and a reference receive cable to connect the fiber optic power meter. In addition to the reference cables, one will need appropriate mating adapters to allow connecting the reference cables to each other and the cable under test. An OLTS will also require reference test cables for each instrument.

OTDRs also use reference cables to connect to the cable under test, but much longer ones due to the different test method and its requirements. This will be covered in the discussions about OTDRs and OTDR testing.

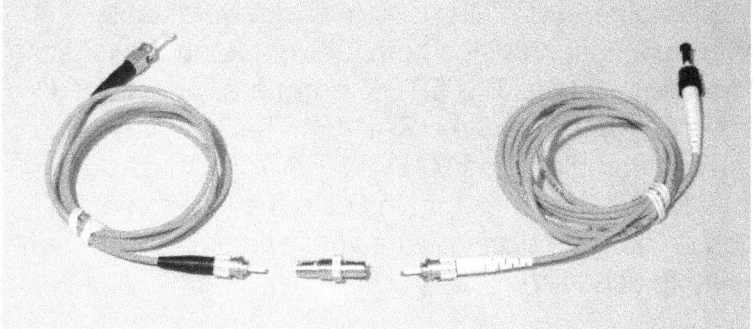

Reference cables and mating adapter for insertion loss testing

For accurate measurements, the launch and receive cables must be made with fiber and connectors matching the cables to be tested and terminated carefully to ensure low loss at the connections. To provide reliable measurements, launch and receive cables must be in good condition and the connectors kept very clean, generally cleaning before each measurement.

Some test standards call for "reference quality" cables with much lower loss connections. There may be restrictions on using bend-insensitive fibers in reference cables which will be discussed in Chapter 6.

Reference cables should be tested against each other to insure their condition is good. This should be done whenever setting a 0 dB reference or if the cables show damage when inspected. Bad reference cables will also show higher loss over time due to wear so the tech should look for gradual changes in tests toward higher loss.

Connector mating adapters are used to connect the cables under test to the launch and receive cables. Only the highest performance mating adapters should be used, and their condition checked regularly, since they are vitally important in obtaining low loss connections. Inexpensive mating adapters may use plastic alignment sleeves that will wear out quickly and leave dirt on connector ferrules. Most adapters use metal alignment sleeves that will last for hundreds of tests. The more expensive mating adapters use ceramic alignment sleeves that will last for many hundreds or even thousands of tests.

For multimode fiber testing, the launch cable will generally use some form of modal control, usually a mandrel wrap, according to test standards. Singlemode launch cables need only a single small loop mode filter. This will be covered more in Chapter 6.

Optical Time Domain Reflectometer (OTDR)

The optical time domain reflectometer (OTDR) is an extremely valuable instrument for testing fiber because it provides both loss data and the location of the loss, essential for troubleshooting. The OTDR is generally described as the fiber optic test instrument that works like RADAR. That's a fairly good analogy but the OTDR is more complex than RADAR. RADAR sends out a radio wave pulse and looks for echoes back that indicate the location of an airplane. The OTDR sends an optical pulse down a fiber and gets much more information back than a simple echo like RADAR.

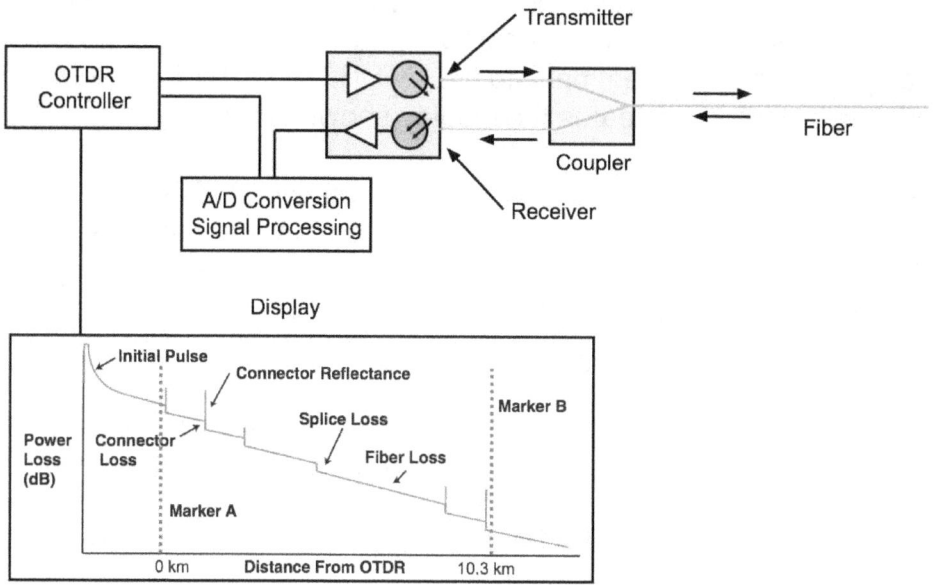

OTDR block diagram

The OTDR has a high power laser transmitter that sends a signal out to the fiber through a bidirectional coupler. As the light travels down the fiber, a small amount of the light is scattered back toward the OTDR and larger amounts are reflected from reflective joints like connectors.

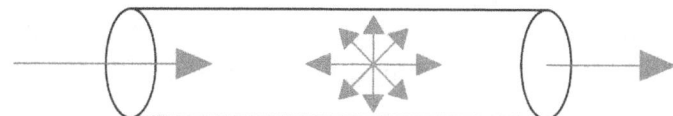

Scattering in an optical fiber

The OTDR uses the phenomena of fiber backscatter to characterize fibers and installed cables, find faults and optimize splices. Scattering is one of the primary loss factors in fiber (the other being absorption), so the OTDR can send out into the fiber a high powered laser pulse and measure the light scattered back toward the instrument.

The backscattered light is very low power, only about 1 millionth of the power in the test pulse so the OTDR needs to average many tests to reduce the noise inherent in a single measurement. This averaging is done digitally after the return signal is digitized.

The OTDR test pulse is attenuated on the outbound leg and the backscattered light is attenuated on the return leg, so the returned signal can be analyzed to determine loss. As the test signal passes through a connector or splice, the loss of that component also reduces the return power allowing the measurement of that loss. If a component is reflective like most connectors, a higher amount of power compared to the backscatter signal results in a light pulse being sent back to the OTDR. The OTDR measures the optical power over time and creates the display shown above.

Light travels in the fiber at a speed determined by the index of refraction of the fiber. Since the index of refraction of the glass (N) is about 1.46, the speed of light in the fiber is calculated as C/N where C is the speed of light. By timing the return signal (T) and knowing the speed of light in the fiber, the OTDR can calculate the distance at any time by T times C/N.

Thus the OTDR can take the data from measuring the power over time and convert time to distance, creating the OTDR trace of power loss over distance. This allows the trace to be analyzed to measure fiber loss and the loss of individual components in the fiber as well as also locate the source of the loss and the distance from the instrument.

Using the backscatter and reflective signals, the OTDR can be used to locate fiber breaks, splices and connectors as well as measure loss. In addition, the OTDR gives a graphic display of the fiber being tested. The OTDR offers a major advantage over the source/FO power meter or OLTS, in that it requires access to only one end of the fiber.

This complex instrument creates a display, called a "fiber trace" or "signature" that is a snapshot of the fiber under test. The drawing below shows how the fiber trace shows details of the fiber corresponding to the location of features on the fiber itself. These features are called "events." Markers, shown by the vertical dashed lines, are used to identify events on the trace and set limits for making measurements.

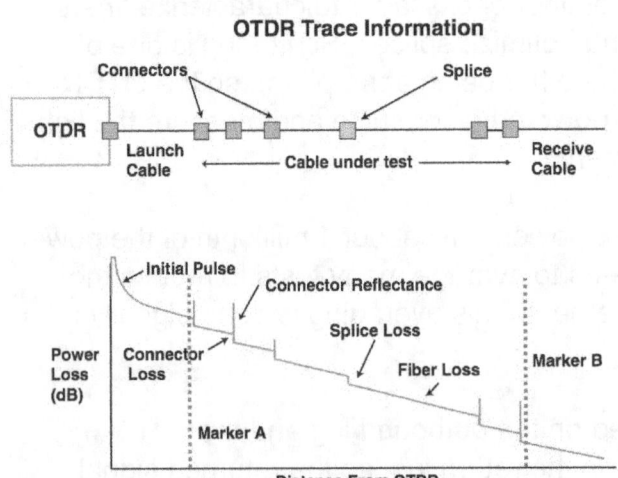

The OTDR trace shows events corresponding to the actual fiber features

The high power of the OTDR test pulse generally overloads the receiver due to reflection in the OTDR connectors and crosstalk in the coupler. This causes the initial pulse in the trace above to take longer to recover creating what is called a "dead zone" near the instrument. A long launch cable, sometimes called a pulse suppressor cable, is used to allow the OTDR to recover before any measurements are attempted. The launch cable has a second and also important use; it allows the OTDR to measure the loss of the first connector on the cable under test.

OTDRs should never be used without a launch cable unless the OTDR is only being used to measure distance or find faults or breaks in the fiber. Measuring the loss of the fiber will require a launch cable and it should also have a receive cable on the far end to allow measuring the final connector on the cable.

Because the OTDR uses backscatter to indirectly measure loss in the fiber, the test results are different from insertion loss testing with a test source and power meter. The test source and power meter measures the loss of the fiber in the same way a fiber optic transmission system would use the fiber instead of the indirect method used by the OTDR. Due to this discrepancy in measurement technique, all standards call for insertion loss testing with a test source and power meter first. OTDR testing is a secondary test but often required on long outside plant cables because only an OTDR can verify the loss of splices and find other sources of loss such as stress on the cable that causes bending loss.

OTDRs must also be matched to the fibers being tested in both wavelength and fiber core diameter to provide accurate measurements. Thus many OTDRs have modular sources to allow substituting a proper source for the application.

The uncertainty of the OTDR measurement is heavily dependent on the backscatter coefficient, which is a function of intrinsic fiber scattering characteristics including core diameter (mode field diameter in singlemode fiber) and numerical aperture. It is the variation in backscatter coefficient that causes many splices to show a "gain" instead of the actual loss.

Another issue with the OTDR is the interrelated issues of distance range and event resolution. To get measurement over long distances, the OTDR needs a very powerful test pulse. Since the output power of the laser is limited, the OTDR sends out a longer test pulse which provides the additional power. However, another limitation of the OTDR is distance resolution. The test pulse has a pulse length that limits the resolution of the instrument because two closely spaced events may be within the span of the test pulse. To get longer range, the pulse must be longer but that means less resolution of individual events. To get more resolution, the pulse must be shorter but that limits the range. Additional range can be attained by averaging the test signals more times but that extends test time. Generally the operator chooses a compromise between range and resolution.

Because of the number of test parameters that can be chosen on an OTDR, the setup of the instrument for a given test situation is very important. That will be discussed in Chapter 7 on OTDR tests. OTDRs also generally have an automatic test function that chooses setup parameters automatically. Most of these auto test functions have to make compromises, e.g. range vs. distance, so using auto test requires careful monitoring to ensure the instrument is not losing valuable
data in a compromised setup.

Most OTDRs today use a common PC operating system on which proprietary software controls the instrument and processes the measurements. Each OTDR's software is somewhat different. In talking to manufacturers, reading manuals and discussing OTDR operating methods with very experienced users, it seems that each instrument has slightly different systems. In evaluating OTDR performance, it is generally impossible to separate the effects of hardware and software operation.

In addition, OTDRs are very complex instruments. It can take days of work to understand how to use some functions, just like most other PC software. We highly recommend using the manufacturer's manuals to learn how to use your instrument properly and also use this book as a way of understanding how the OTDR can make measurements to ensure getting the most out of such a complex instrument.

While most OTDR applications involve finding faults in installed cables or verifying splices, they are very useful in manufacturing for inspecting fibers for manufacturing faults. Development work on improving the short range resolution of OTDRs for LAN applications and new applications such as evaluating connector return loss promise to enhance the usefulness of the instrument in the future.

OTDRs come in several basic versions. Full size or laboratory OTDRs offer the highest performance and have a full complement of features but are very big and high priced. Mini OTDRs provide the same type of measurements as a full OTDR, but with fewer features to trim the size and cost for field use. Fault finders use the OTDR technique, but greatly simplified. Most just provide the distance to a fault, to make the instruments more affordable and easier to use. There are even OTDR modules without a display that use a PC or tablet for readout.

Specifications For OTDRs
Specifications for OTDRs include the optical measurement parameters such as wavelength, dynamic range and measurement accuracy. In addition, the operational parameters that allow setting up the instrument for specific fiber tests such as distance range, pulse widths and averaging options should be included. Specifications shows are for example only.

Optical Sources
      Singlemode module: 1310 ± 25 nm, 1550 ± 25 nm
      Multimode module: 850 ± 30 nm, 1300 ± 30 nm
Distance ranges: 0.2 to 512 km
Pulse width: 3 to 20,000 nanoseconds (ns)
Event dead zone (minimum): 0.6 m
Attenuation dead zone: 4 m
Dynamic range: 40 dB
Distance measurement accuracy: 0.02% distance ± 1m
Loss measurement accuracy: ± 0.05 dB/dB
Reflectance measurement accuracy: ± 2 dB
OTDRs often have power meter and test source options which will be specified as would be any power meter or test source.

Fiber Identifiers

Technicians often need to identify a fiber in a splice closure or at a patch panel. If one carefully bends a singlemode fiber enough to cause loss, the light that couples out can also be detected by a large area detector. A fiber identifier uses this technique to detect a signal in the fiber at normal transmission wavelengths.

These instruments usually function as receivers, able to discriminate between no signal, a high speed signal and a 2 kHz tone. By specifically looking for a 2 kHz "tone" from a test source coupled into the fiber, the instrument can identify a specific fiber in a large multifiber cable, especially useful to speed up the splicing or restoration process.

Fiber identifiers can be used with both buffered fiber and jacketed single fiber cable. With buffered fiber, one must be very careful to not damage the fiber, as any excess stress here could result in stress cracks in the fiber which could cause a failure in the fiber anytime in the future.

Instruments For Measuring Dispersion (Fiber Bandwidth)

Although fiber has a very high bandwidth, some applications today actually approach its limits, requiring performance evaluation. Two factors limit multimode fiber bandwidth: modal dispersion and chromatic dispersion. Specialized instruments are available for testing each of these specifications but are expensive and never used outside the laboratory. Fast networks generally specify maximum lengths for multimode fiber of various grades (OM2, OM3, OM4 or OM5) instead of requiring field testing

Long singlemode links require concern over chromatic dispersion (CD) or polarization-mode dispersion (PMD). The long singlemode fiber optic cable plants that were installed years ago need testing to determine if they are capable of upgrades. New singlemode cable plants of longer lengths are tested to determine what their capability is for future upgrades. This process, called "fiber characterization" and its relevant test equipment will be covered in Chapter 8.

O/E and E/O Converters

Optical to electrical (O/E) and electrical to optical (E/O) converters are used in the laboratory for viewing fiber optic waveforms and testing fiber bandwidth. O/E converters can be used with high speed oscilloscopes to analyze pulses in fiber optic links to see if the waveforms are of the proper shape. This means measuring rise and fall times of the pulse and the depth of modulation (the difference between the peak power of the pulse and the lowest power reached between pulses. They can be used for testing lasers and LEDs used in transmitters and link dispersion in long links. E/O converters are used to test receivers for bandwidth and margin, usually in conjunction with a bit error rate tester and attenuator.

Optical Continuous Wave Reflectometer (OCWR)

The OCWR or reflectance tester was originally proposed as a special purpose instrument to measure the reflectance or optical return loss of connectors installed on patchcords or jumpers. Unfortunately, its purpose became muddled between conception and inception.

As actual instruments came on the market, they had much higher measurement resolution than appropriate for the measurement uncertainty (0.01 dB resolution vs. 1 dB uncertainty), leading to much confusion on the part of users as to why measurements were not reproducible. In addition, these instruments were touted as a way to measure the optical return loss of an installed cable plant, integrating the backscatter of the fiber with any reflections from connectors or splices.

Since an OTDR is the only way to test actual connections in installed cable plants for reflectance and they can be programmed to also read return loss, the OCWR has seen little use in fiber optic testing.

Optical Fiber Analyzers

There are many parameters of optical fiber that require testing by the manufacturer following production of the fiber. These include attenuation (as a function of source wavelength), bandwidth/dispersion, numerical aperture and all the physical dimensions such as core and cladding diameter, ovality, and concentricity. Automated laboratory instruments are available to measure all these parameters and more automatically, but some fiber manufacturers prefer to build their own.

Remote Fiber Test Systems

Remote fiber test systems are used to monitor installed fiber optic cable plants during operation. They may work by monitoring the insertion loss of dark fibers, tapping live fibers to monitor loss by measuring the signals at transmitter at receiver ends. Some use OTDRs and inline filters to test at out of band wavelengths. These instruments are not common since many system transceivers today have power monitoring functions.

Fiber Optic Talksets

While technically not a measuring instrument, fiber optic talksets are useful for installation and testing. They transmit voice over fiber optic cables already installed, allowing technicians splicing or testing the fiber to communicate effectively. Talksets are especially useful when walkie-talkies and cell phones are not available, such as in remote locations where splicing is being done in buildings where radio waves will not penetrate or in secure facilities where radio communications are not allowed.

Attenuators

Fiber optic attenuators may not seem like test equipment but are highly useful tools for network simulation and testing. Attenuators are used in the laboratory to simulate the loss of long fiber runs for testing link power margin in network simulation tests. In the field, attenuators are used to stress test links and reduce power levels when power is too high at the receiver. Attenuators are also used for self-testing links in a loopback configuration.

In margin testing, variable attenuators are used to increase loss until the system has a high bit error rate. For loopback testing, an attenuator is used between a single piece of equipment's transmitter and receiver to test for operation under maximum specified fiber loss. If systems work in loopback testing, they should work with a proper cable plant. Thus many manufacturers of network equipment specify a loopback test as a diagnostic/troubleshooting procedure.

Attenuators have two relevant specifications, loss and reflectance. They can be made by gap loss, or a physical separation of the ends of the fibers, inducing bending losses or inserting calibrated optical filters. Both variable and fixed attenuators are available, but variable attenuators are usually used for testing. Fixed attenuators may be inserted in the system cables where distances in the fiber optic link are too short and excess power at the receiver causes transmission problems.

Chapter Exercises

- Exercises using these instruments will be included in the following chapters about use of each of them.
- Look at manufacturer's websites for information on these instruments. Search for the name of the instrument and see who manufactures them. Look for application notes and instructions you can download.

Chapter Quiz

1. Dirt and contamination is the most common problem encountered in fiber optics.
    True
    False

2. To effectively inspect a fiber optic connector, a microscope must be at least 400 power.
    True
    False

3. For multimode fiber testing, the launch cable will generally use some form of modal control, usually a mandrel wrap, according to test standards.

     True

     False

4. OTDR testing will generally not give the same test results as a light source and power meter.

     True

     False

5. What fiber optic instrument creates a "snapshot" of the fiber under test?

     A. Inspection microscope

     B. OLTS

     C. OTDR

     D. Visual fault locator

6. Microscope inspection of the connector ferrule allows one to find_____.

     A. Scratches in the fiber

     B. Cracks in the fiber

     C. Dirt or contamination on the end of the ferrule

     D. All of the above

7. Visual fault locators with a laser source can trace OSP singlemode fibers for lengths of _____ km.

     A. 0.5-1

     B. 2-5

     C. 5-20

     D. 20-100

8. Insertion loss testing is done with a _____.

     A. Light source and power meter

     B. Light source and power meter and reference test cables

     C. OTDR with a launch reference cable

     D. OTDR with two reference test cables

9. In order to measure the loss of the connector on the far end of a fiber with an OTDR, you must _____.

     A. Attach a reference cable to the far end of the fiber being tested

     B. Test a second time with the OTDR at the far end of the cable

     C. Either A or B

     D. None of the above, the OTDR can not make that measurement

10 Reference test cables for either insertion loss or OTDR testing must
_____.
   A. Have the same fiber type as the cable plant being tested
   B. Have connectors that can be mated to those on the cable plant
   C. Have connectors in good condition
   D. All of the above

11. The reference cables you need for testing _____.
   A. Can be any old cables in your toolkit
   B. Should be random patch cables used for connecting equipment to the
      cable plant
   C. Should be known good cables regularly tested for low loss
   D. Must be special reference-grade test cables which can only be
      purchased from the test equipment manufacturer.

Chapter 5

Visual Inspection and Cleaning Of Connectors

Objectives: From this chapter you should learn:
Why it is important to inspect and clean fiber optic connectors
How should connectors be inspected for cleanliness
What processes are used for cleaning fiber optic connectors
How to prevent problems with connectors

Introduction

Dirty connectors are one of the major problems in fiber optics, causing high connector loss, high reflectance and contaminating transceivers. Network operators claim that 15-50% of all network problems can be traced to dirty connectors causing connection problems.

One of the first visits we made to a fiber optic network user's facilities to demonstrate test equipment, we watched as a technician explained that fiber was so small that it was very sensitive to dirt on the connector. Every connector needed cleaning, he explained, and he demonstrated how he did it by wiping it on his shirt several times with a circular motion.

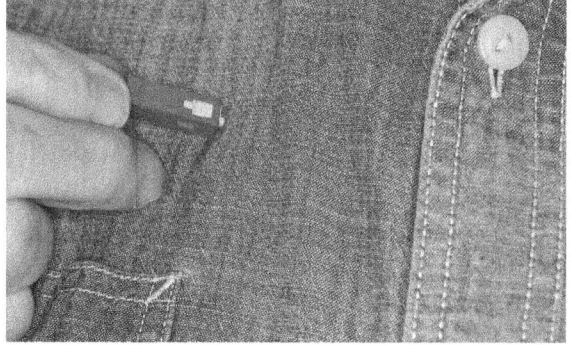

While his intentions were good, his methods, at least with our current knowledge, was not. The shirt was neither clean or lint free. It probably added more contamination to the connector than it removed.

Typical environmental dirt is relatively large compared to the size of the core of a singlemode fiber. Much of the dirt is silica-based and hard enough to scratch the fiber if sandwiched between two spring-loaded ferrules. Some cleaning processes may cause problems if done incorrectly; adding a film to the end of the ferrule or causing static electricity that attracts more dirt.

And then there is the issue of "dust caps." Within the fiber optic industry, some say that the are called dust caps because they may be filled with dust. Dust caps are molded by the millions, dumped into buckets by the molding machine, packaged and shipped to connector manufacturers in large quantities. They have mold release on them that can attract and hold dirt. At every step along the process they are exposed to environmental dust and contamination.

These plastic caps for ferrules or connector bodies should be called "protective caps" because they do protect connectors from damage, for example when dropped, or from additional contamination from touching objects or people's skin. It is not advisable to clean a connector, put on a protective cap and assume the connector will stay clean. When you take them off, clean and inspect the connector to ensure its is clean.

Optical Inspection Of Connectors with Microscopes

There are two major uses for visual inspection of fiber optic connectors. Polished connector ferrules require visual inspection during manufacturing to evaluate polishing and find possible defects during the connector termination process. In the field, connectors need inspection for cleanliness and damage before testing or connection to another connector or transceiver.

Visual inspection is accomplished using a microscope that has a fixture to hold the fiber or connector steady in the field of view and a light source to illuminate the connector.

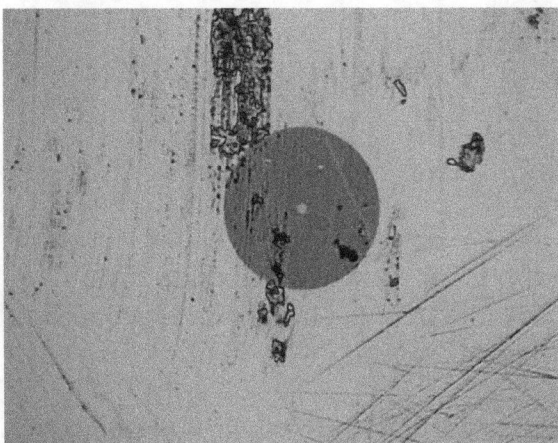

Optical image of connector ferrule end showing dirt and scratches

Very inexpensive microscopes included in termination kits are usually modified 100X plastic microscopes intended for science student lab use with an adapter for fiber optic connector ferrules, primarily 2.5 or 1.25 mm.

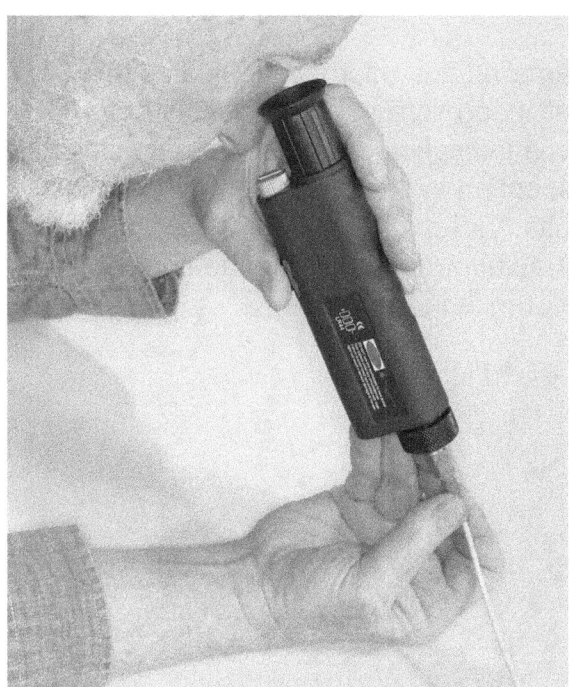

Portable optical microscope used for connector inspection

Microscopes designed specifically for fiber optics inspection have more precise connector adapters and usually include filters to protect the user from infrared light that might be present in communications systems. These microscopes also offer higher magnification, 100X to 400X, for closer inspection of polished ferrule ends. Most have better lighting, often direct down the axis of the microscope to see the end clearly and at an angle which helps diagnose polishing problems.

Note On Eye Safety: Visual microscopes should have a built-in infrared filter to remove any signals being transmitted in the fiber for the protection of the eye of the user. An optical microscope can capture any light in the fiber and focus all of it into the eye, a potential danger to the user.

Since the light in most fiber systems is in the infrared (IR) and invisible to humans, it will not be detected visually, even if the power level is high enough to be dangerous. Most fiber optic systems have power levels too low to be harmful but some might - especially telecom and CATV systems with fiber amplifiers or WDM.

One should always check power levels with a power meter before inspecting connectors with a microscope. If possible, only use microscopes with IR filters to prevent IR light from entering the eye. A video microscope is preferable since it offers no danger to the eye.

Video microscopes use small video cameras and microscope lenses to provide a display of the view on a small video screen or the display of a PC or tablet. They offer more flexibility in magnification and image manipulation. Many of these also offer automatic inspection to international standards for cleanliness and produce pass/fail results. Video microscopes also allow for storing an image of the connector, valuable for documenting the condition of the connector at installation and for reference in the future.

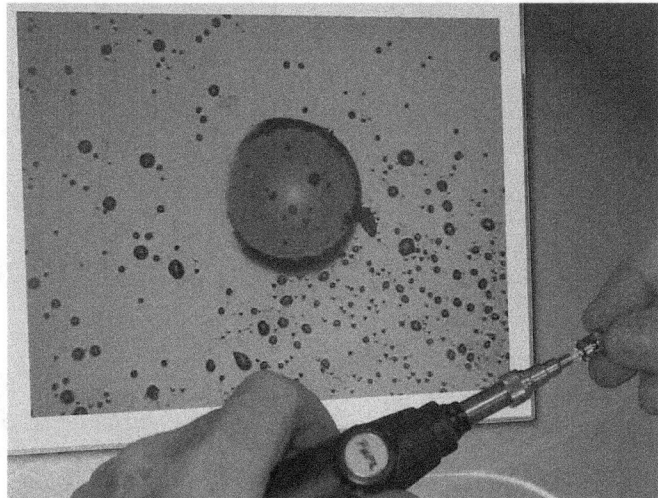

Video microscope view of dirty connector

Fiber optic inspection microscopes vary in magnification from 30 to 800 power, with 100-400 power being the most widely used range for connector ferrule inspection. Higher magnification is helpful when for inspecting for proper polish and scratches where you are looking for micron-sized defects. Lower magnification is used to view more of the connector when looking for dirt or contamination.

New wide field video microscopes allow looking at the whole ferrule and inside the connector body looking for contamination. These are very useful tools for inspecting for cleanliness, since they allow you to see more of the connector ferrule and body where dust can accumulate and then migrate back to the connector ferrule end where it becomes a problem when mating connectors.

Wide field video microscope view of connector

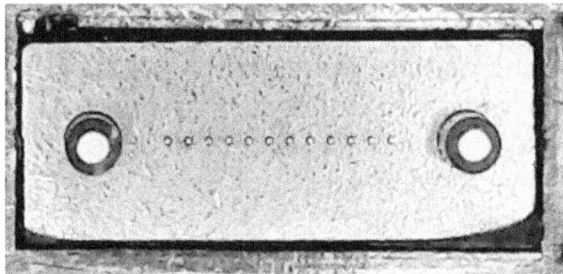

Wide field microscopes are also useful for inspecting MPO connectors

Lighting is usually provided two ways, direct along the axis of the connector ferrule and at an angle to the ferrule end. Direct lighting helps inspect fibers for cracks and chips or to see if the fiber is aligned properly in the ferrule hole. Angle lighting is used to more clearly show the polish quality and scratches since it shows shadows of any surface irregularities.

Inspecting Connectors During Termination

When inspecting connectors during termination, you will see lots of variation in end finishes depending on the quality of the termination. Below is something rarely seen, a perfect connector ferrule end face. It's so perfect it doesn't look like a photograph but instead a drawing.

A perfect polished and clean connector ferrule

Inspecting Connectors For Proper Termination

Connectors are always inspected during the termination process to ensure termination has been done properly. Factory terminations are machine polished and usually inspected by video microscopes that automatically look for defects and provide a report on the connector. These reports may be provided with the finished cable as quality control data.

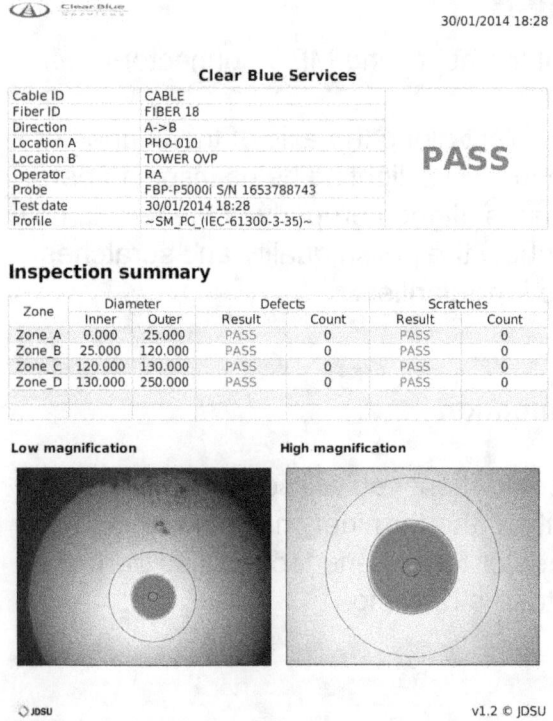

Clear Blue	30/01/2014 18:28

Clear Blue Services

Cable ID	CABLE	
Fiber ID	FIBER 18	
Direction	A->B	
Location A	PHO-010	**PASS**
Location B	TOWER OVP	
Operator	RA	
Probe	FBP-P5000i S/N 1653788743	
Test date	30/01/2014 18:28	
Profile	~SM_PC_(IEC-61300-3-35)~	

Inspection summary

Zone	Diameter		Defects		Scratches	
	Inner	Outer	Result	Count	Result	Count
Zone_A	0.000	25.000	PASS	0	PASS	0
Zone_B	25.000	120.000	PASS	0	PASS	0
Zone_C	120.000	130.000	PASS	0	PASS	0
Zone_D	130.000	250.000	PASS	0	PASS	0

Low magnification   High magnification

JDSU   v1.2 © JDSU

Automated fiber inspection report from video microscope

Factory made singlemode connectors may also have inspection done by an interferometer, a device that can provide a profile of the end of the connector. It can verify the end radius and its conformance to standards. It can also profile the polished fiber end, show how well it fits the profile of the connector ferrule, and find the center of the radius of the curvature of the end of the polished fiber.

Interferometer profile of polished fiber in connector ferrule (FIBO)

Field terminations are more likely inspected with a visual microscope included in a termination kit and the pass/fail decision is made by the judgment of the technician. An experienced technician knows what are the limitations for scratches or defects on the connector based on experience in termination and testing many connectors. The photo at the beginning of this section showed what a excellent connector looks like, but not all connectors are this good. If you look at lots of field-terminated connectors, you are likely to see some ferrules like these.

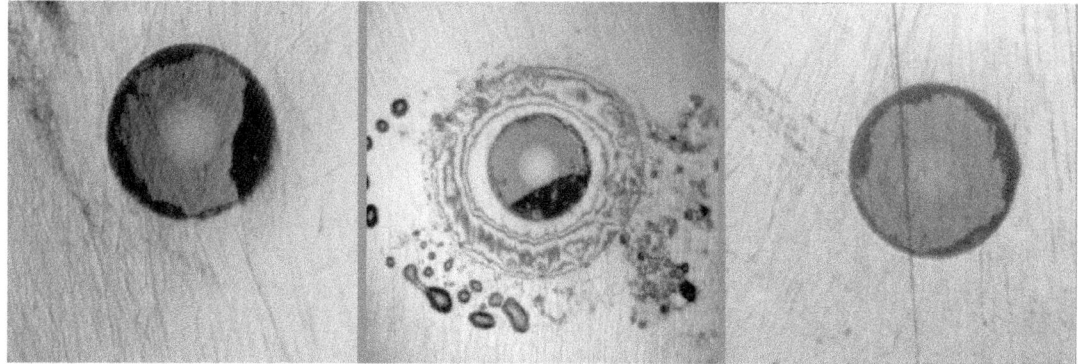

Microscope views of 3 bad connectors

These three connectors show some serious problems in termination. The connector on the left has a large chip on one side and small chips all around. It is heavily scratched and has some pitting. The connector in the middle shows contamination from some liquid, but the rings around the fiber show the epoxy on the fiber end was not completely polished off. The connector on the right shows scratching and chipping also, including one very large scratch across the fiber and the ferrule. All are rejects.

Field Inspection Of Connectors For Cleanliness

In the field, the biggest use for microscope inspection of connectors is to determine if the connectors are clean enough to use. For cleanliness inspection, lower magnification is preferred because it allows viewing more of the end of the connector ferrule.

Here are some photos of connectors needing to be cleaned.

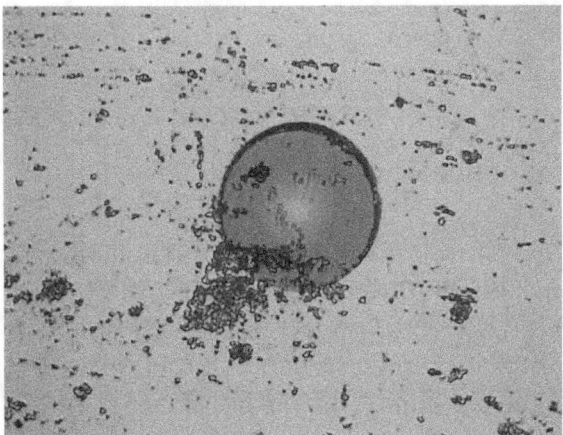

Connector with large amounts of surface dirt

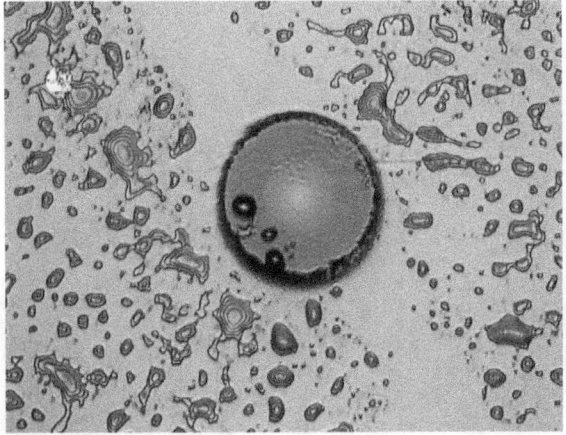

Connector contaminated by oily residue

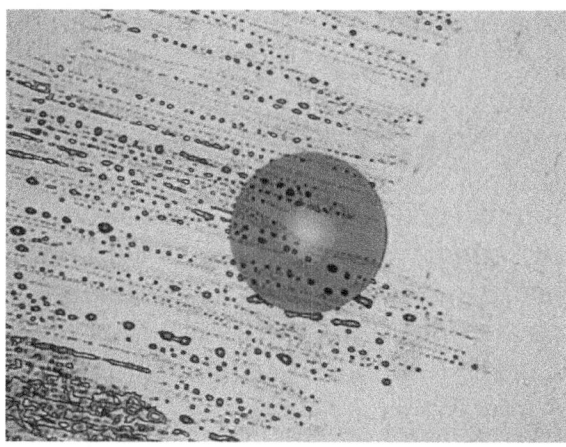

Connector smudged by oil on fingertip

Many connectors will have both dirt and residue on the ends that need cleaning. The type(s) of contamination and dirt will only be known if the connector is inspected visually.

Cleaning Fiber Optic Connectors

Going all the way back to the technician we mentioned cleaning a connector on his shirt, the recommended procedure was to use reagent-grade 99% isopropyl alcohol and a lint free wipe. Isopropyl alcohol (IPA) was recommended because it was an effective solvent to remove most oily contamination and was not harmful to the epoxies used in termination. However it was hard to find and if bought in bulk, it absorbed water from the air (hygroscopic) and became easily contaminated.

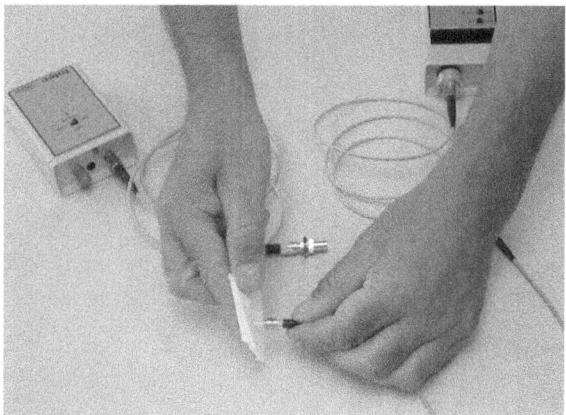

Alco-Pads being used to clean connector for testing

A better solution was to buy prepackaged lint-free alcohol-soaked pads called "Alco-Pads." The sealed packages kept the alcohol pure and proper use of the pads made for easy "wet-dry" cleaning. The recommended method was to open one corner of the sealed package and start cleaning on the exposed pad while it was wet. The alcohol evaporated quickly leaving dry sections of the pad, so the two steps of cleaning were to clean on a wet section and dry on a dry section of the pad. This generally produced acceptable results.

Often you will see techs with a tabletop bottle of IPA and some wipes. The IPA rapidly becomes contaminated by moisture and wipes will attract dirt or pick dirt up from the tabletop. Even in patchcord manufacturing facilities, these processes may be used and connectors are often only superficially clean. Always remember to inspect and clean brand new patchcords out of their sealed bags before using them.

As the need for proper cleaning of fiber optic connectors became better known, manufacturers of cleaning products began to do research on how to clean connectors properly and created products aimed at special applications. They developed dry connector cleaners, using treated lint-free tapes in cassettes, boxes or small hand-held tools (probes) that could reach into a mating adapter and clean a connector at the far end. Some of the probes can even clean the mating adapter, although that is usually best done with a special swab.

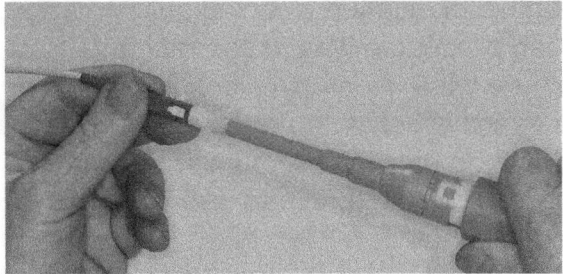

Probe type cleaner used with SC connector

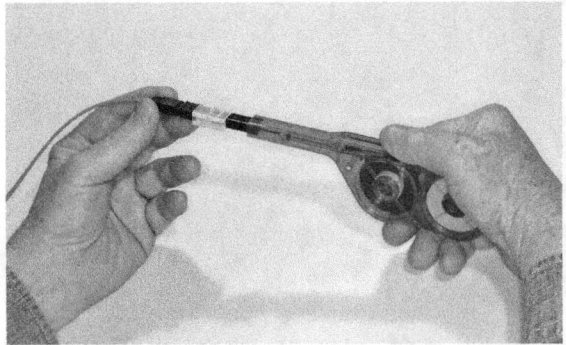

Probe type cleaner used on wide ferrule multifiber MPO connector

These dry cleaners are very convenient and fairly effective. One drawback of the dry cleaners is they may generate a static charge on the end of the fiber optic connector ferrule that attracts airborne dust. Rubbing a glass rod with a silk cloth is a classic method of demonstrating static electricity. Some of the dry cleaners now have been designed to be conductive to prevent static buildup.

Wet/dry cleaning is generally the preferred method of cleaning connectors. The wet part of the process will loosen dirt and contamination and the dry process will remove them. Much research has gone into developing cleaning solutions that are better at removing dirt, do not generate static and are safer. IPA and many cleaners, including that in typical "canned air" are flammable! They are not safe for use in all environments (remember fusion splicers use an electric arc) and often cannot be legally shipped.

The wet/dry process works with a fairly large wipe or the boxes of cleaning wipes shown below. You wet on end of the cleaning wipe with a cleaner solution then wipe the connector from the wet to the dry side of the wipe. This quick process should get the connector clean but only visual inspection will confirm the connector has been properly cleaned.

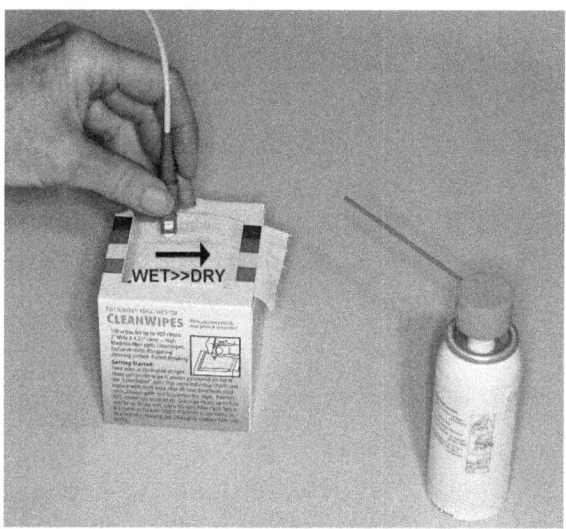

Wet to dry cleaning in one swipe

Protecting And Cleaning Test Equipment And Cables

While we have been mostly discussing cleaning cables, a process that applies to any cable, the fiber tech also needs to apply these cleaning guidelines to their reference test cables and equipment to ensure that dirt and contamination do not adversely affect testing.

Obviously, reference test cables need special handling. They should be cleaned before use and every cable they test must be cleaned before use.

Mating a perfectly clean reference cable to a dirty connector on a cable under test will likely not only cause high loss but the dirty connector may contaminate the connector on the reference cable and the mating adapter used. Consistent cleaning is very important.

The same advice about cleaning cables apples to test equipment. The connector receptacles on test equipment should be cleaned in the same manner as transceivers. Dirt here will affect the output power of the test source and input of the meter, and changes in the cleanliness will change the calibration of test setups.

Of course sources should be cleaned before attaching a (cleaned) connector on a launch reference cable. Of course that cable should not be disconnected or the 0 dB reference will be lost and have to be recalibrated.

Meters that have removable adapters on the detector should be cleaned periodically. This can be done by removing the adapter and cleaning it like a mating adapter, then wet/dry cleaning the surface of the detector on the meter with connector cleaning pads.

Both source outputs and meter inputs should have protective caps to keep dust from entering the receptacles when cables are not attached.

The same advice holds for connector inspection microscopes. Imagine how dirty they must get since dirty connectors are inserted into their adapters all the time.

Chapter Exercises

- Using a patchcord with a good connector, remove the dust cap and examine the condition of the ferrule end. Clean and reexamine the connector ferrule. Is it in better condition?
- Touch the end of a clean ferrule with your finger or your face (oily skin is preferable) and inspect the condition of the end of the ferrule. Clean and reexamine the connector ferrule.
- Wipe a clean connector across a rug on the floor and inspect the condition of the end of the ferrule. Clean and reexamine the connector ferrule.
- Using a dirty connector made by one of the processes noted above, try dry cleaning and wet/dry cleaning. Which works best?

Chapter Quiz

1. Problems with fiber optic connectors in systems are usually caused by dirty or contaminated connectors.
> True
> False

2. Dust caps used to protect fiber optic connectors are often dusty, so connectors must be cleaned after the dust cap is removed.
> True
> False

3. Visual inspection of fiber optic connectors is generally done with _____.
> A. Your naked eye
> B. A jeweler's loupe
> C. An optical microscope
> D. An electron microscope

4. Typical magnification used to inspect connectors is _____.
> A. 10-100X
> B. 100-400X
> C. 400-1000X
> D. Magnification doesn't matter

5. Visual inspection of the connector endface with a microscope is used to find_____.
> A. Poor polish or scratches on the ferrule endface
> B. Dirt
> C. Contamination
> D. All of the above

6. Always clean fiber optic connectors _____.
> A. Before connecting patchcords to equipment
> B. Before connecting patchcords to patch panels
> C. Before connecting patchcords to test equipment
> D. All of the above

7. Fiber optic connectors should be cleaned with _____.
> A. Special fiber cleaners or lint free wipes with pure isopropyl alcohol
> B. Cotton pads and distilled water
> C. Canned air and tissues
> D. A wipe on your shirt

8. The best way to clean connectors is the _____ method.
   A. Wet
   B. Dry
   C. Wet to dry
   D. Canned air

9. When inspecting connectors in an operational fiber optic network, it is important to _____ before inspecting a connector.
   A. Check for power in the fiber with a power meter
   B. Clean your microscope objective
   C. Have cleaning supplies ready to use
   D. Find a dust cap to put on the connector

10. International standards call for inspecting the _____ of the connector.
   A. Area just around the fiber itself
   B. The entire ferrule end
   C. The ferrule end and sides
   D. The entire connector

11. The proper process for preparing connectors for connection to another connector or an active device is _____ before connecting.
   A. Clean
   B. Inspect then clean
   C. Inspect, clean and inspect again
   D. Inspect, clean and inspect again until the connector is perfectly clean

Chapter 6

Loss Budgets

Objectives: From this chapter you should learn:
What is link loss
What is a loss budget
How do you calculate a loss budget
How do loss budgets depend on the test method used
What is power budget and how does it differ from a loss budget
What is the power penalty

Understanding Power Budgets And Loss Budgets

The terms "power budget" and "loss budget" are often confused. The power budget refers to the amount of loss that a datalink (transmitter to receiver) can tolerate. Sometimes the power budget has both a minimum and maximum value, which means it needs at least a minimum value of loss so that it does not overload the receiver and a maximum value of loss to ensure the receiver has sufficient signal to operate properly.

The loss budget is the amount of loss that a cable plant should have. It is estimated by adding the losses of all the components used in the cable plant to get the total estimated end-to-end loss. Obviously, the two are related. A data link will only operate if the cable plant loss is within the power budget of the link.

Besides determining if the designed cable plant will support the communications equipment, a loss budget can be used to determine the appropriate loss for the cable plant for testing purposes. After installation, test results can be compared to the loss budget to determine if the installation was done correctly.

Communications System Power Budget
All datalinks are limited by the power budget of the link. The power budget is the difference between the output power of the transmitter and the input power requirements of the receiver.

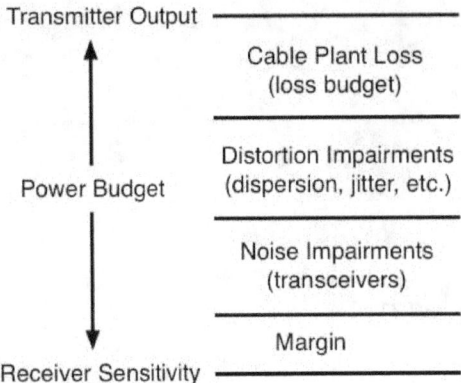

The receiver has an operating range limited on the low power end by the signal-to-noise ratio (S/N) in the receiver. As the power level at the receiver decreases, electrical noise in the receiver can cause signal degradation. The S/N ratio is generally quoted for analog links while the bit-error-rate (BER) is used for digital links. BER is practically an inverse function of S/N.

On the high power end, the limitation for the receiver is the maximum power it can tolerate before overloading, causing distortion and signal degradation. Receivers must operate in the range between where noise and overloading cause signal degradation.

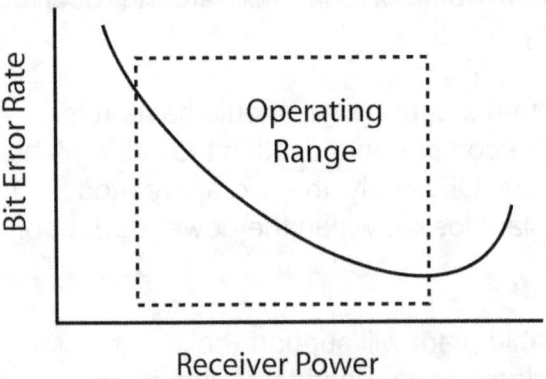

A BER/Receiver power diagram shows the operating range of the receiver.

Transceivers may also be affected by the distortion of the transmitted signal as it goes down the fiber, a big problem with multimode links at high speeds or very long OSP singlemode links.

Cable Plant Link Loss Budget Analysis
Loss budget analysis is the calculation and verification of a fiber optic system's operating characteristics. It is used to estimate the loss of a cable

plant being installed, determine if the cable plant will work with any given transmission system's power budget and provide an estimate for comparison to actual test results.

A link loss budget includes items such as the length of the link, fiber type, wavelengths, connectors and splices, and any other sources of loss in the link. Attenuation and bandwidth are the key parameters for loss budget analysis, but since we cannot test multimode bandwidth, we generally use limits for loss set by the standards for the systems or networks we are going to use on the cable plant. To check link loss, there is a table of link losses from industry standards for many links in the FOA Online Guide. The designer should analyze link loss early in the design stage prior to installing a fiber optic system to make certain the system will work over the proposed cable plant.

From the system standpoint, we have a limit to the loss it can tolerate on the cable plant, called a power budget, determined from the output of the transmitter and the required input of the receiver. We define these errors for the system as "bit-error rate" and they may be caused by too little power or too much power at the receiver. It is important to note that most calculations focus on the cable plant loss being low enough for the system power budget.

However, on some systems, especially laser-based singlemode systems, the receiver may not tolerate a cable plant with very low loss. That may cause very high power at the receiver that may overload it, causing transmission errors. Under such conditions, an attenuator is added at the receiver end of the link to lower the power to an acceptable level.

Both the passive and active components of the circuit can be included in the budget loss calculation. Passive loss is made up of fiber loss, connector loss, and splice loss. Include any couplers or splitters in the link, for example those used in FTTH PON or passive OLAN systems. Active elements like repeaters require the links before and after the repeater be considered as separate links.

The idea of a loss budget is to ensure the network equipment will work over the installed fiber optic link. A big issue is what values you should use for component losses when making the calculations. You can use the values in industry standards like TIA-568 which are considered very conservative (higher loss) for most components. You can use typical values or values from component manufacturers that may be close to typical values. Or the user may have specific values that they specify, not unusual for sophisticated users like telcos. It is normal to be somewhat conservative with the specifications. You probably do not want to use the best possible specifications for fiber attenuation or splice or connector loss to allow some margin for variation in installation and component degradation over time.

Example Of Loss Budget Calculation

The best way to illustrate calculating a loss budget is to show how it's done for a typical cable plant, here a 2 km hybrid multimode/singlemode link with 5 connections (2 connectors at each end and 3 connections at patch panels in the link) and one splice in the middle. See the drawings below of the link layout and the instantaneous power in the link at any point along it's length, scaled exactly to the link drawing above it.

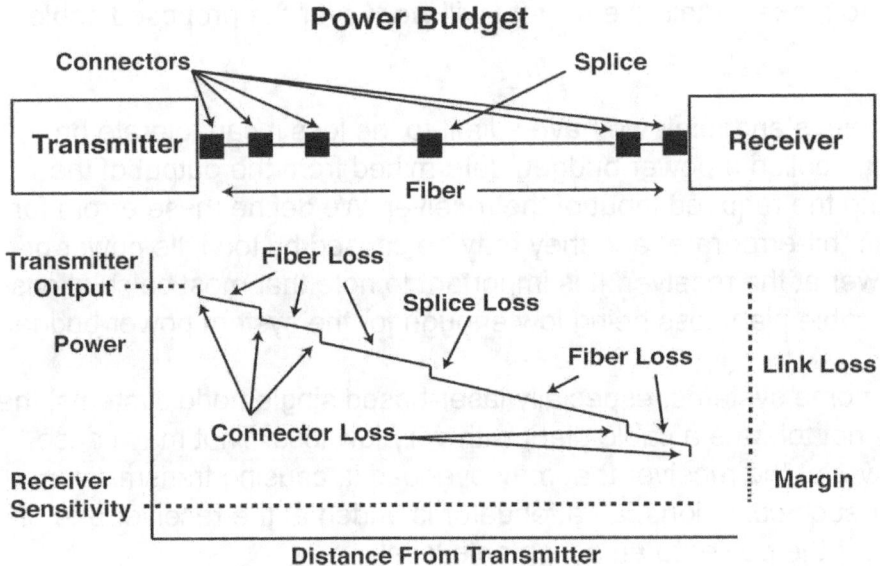

Cable plant loss budget

Step 1. Fiber Attenuation

Calculate the contribution of the optical fiber attenuation at the operating wavelengths (length times standard estimates of loss at each wavelength)

Fiber Attenuation Estimates

Cable Length (km)	2.0	2.0	2.0	2.0
Fiber Type	Multimode		Singlemode	
Wavelength (nm)	850	1300	1310	1550
Fiber Atten. (dB/ km)	3 [3.5]	1 [1.5]	0.4 [1/0.5]	0.3 [1/0.5]
Total Fiber Loss (dB)	6.0 [7.0]	2.0 [3.0]	0.8 [2/1	0.6 [2/1]

Note: All specifications in brackets are maximum values per EIA/TIA 568 standard. For singlemode fiber, a higher loss is allowed for premises applications, 1 dB/km for premises, 0.5 dB/km for outside plant.

Step 2. Connector Loss

Multimode connectors will have losses of 0.2-0.5 dB typically. Singlemode connectors, which are factory made and fusion spliced on will have losses of 0.1-0.2 dB. Field terminated singlemode connectors may have losses as high as 0.5-1.0 dB. Let's calculate it at both typical and worst case values.

Connector Loss	Typical adhesive/polish connector	Prepolished/splice connector. MPO connector or TIA-568 maximum acceptable
	0.3 dB	0.75 dB
Total # of Connectors	5	5
Total Connector Loss	1.5 dB	3.75 dB

All connectors are allowed 0.75 max per EIA/TIA 568 standard.

Note: Many designers and technicians wonder when doing a loss budget whether the connectors on the end of the cable plant should be included in the loss budget. When the cable plant is tested, the reference cables will mate with those connectors and their loss will be included in the measurements. In the chapter on insertion loss, we will discuss how to set a reference for loss using one, two or three reference cables. This not will make more sense after reading that chapter.

If the "0dB" reference for the insertion loss test was done with only one cable and the light source and power meter, the launch reference cable, which is the most common way, the connectors on the end of the cable will be included in the loss so the loss budget should include both connectors.

If the "0dB" reference for the insertion loss test was done with three cables, the launch reference cable, a receive reference cable and a third reference cable between them, a method used for many plug and jack (male/female) connectors such as MPOs, the loss budget should not include the connectors on the end. When making the "0dB" reference with three cables, two connections are included in setting the reference so the measured value will be reduced by the value of those two connections. If the loss budget is calculated without the connectors on the ends, the value will more closely approximate the test results with a 3-cable reference.

While the two-cable reference method is rarely used, it includes only one connector. Thus you could use the same approach when calculating loss budgets for this test method and include only one of the end connectors.

Whatever test method is presumed, it mut be documented when the loss budget is calculated.

Step 3. Splice Loss

Multimode splices are usually made with mechanical splices, although some fusion splicing is used. The larger core and multiple layers make fusion splicing about the same loss as mechanical splicing, but fusion is more reliable in adverse environments. Figure 0.1-0.5 dB for multimode splices, 0.3 being a good average for an experienced installer. Fusion splicing of singlemode fiber will typically have less than 0.05 dB (that's right, less than a tenth of a dB!)

Splice Loss	0.3 dB
Total # splices	1
Total Splice Loss	0.3 dB

(For this loss budget calculation, all splices are allowed 0.3 max per EIA/TIA 568 standard)

Step 4. Total Cable Plant Loss

Add together the fiber, connector and splice losses to get the total link loss of the cable plant.

	Multimode		Singlemode	
	Best Case [TIA 568 Max]		Best Case [TIA 568 Max, premises/OSP]	
Wavelength (nm)	850	1300	1310	1550
Total Fiber Loss (dB)	6.0 [7.0]	2.0 [3.0]	0.8 [2/1]	0.6 [2/1]
Total Connector Loss (dB)	1.5 [3.75]	1.5 [3.75]	1.5 [3.75]	1.5 [3.75]
Total Splice Loss (dB)	0.3 [0.3]	0.3 [0.3]	0.3 [0.3]	0.3 [0.3]

Other (dB)	0	0	0	0
Total Link Loss (dB)	7.8 [11.05]	3.8 [7.05]	2.6 [6.05/5.05]	2.4 [6.05/5.05]

These values of cable plant loss should be the criteria for testing. Allow +/- 0.2 -0.5 dB for measurement uncertainty and these become your pass/fail criteria.

Note: Part of the reason for calculating loss budgets is to have a "pass/ fail" limit for testing an installed cable plant. When doing so, one needs to remember that the loss budget is an estimate, based on some guidelines for component losses. In addition, the test results have measurement uncertainty. When measuring the loss of a cable plant and comparing the data to the loss budget, one should not forget these variables. Exceeding a loss budget estimate by 0.1dB, for instance, is hardly reason to fail a fiber in a cable plant. If the loss is more than 10% high, e.g. 0.5dB on a 5dB loss budget, it is cause for concern and justifies some troubleshooting.

Equipment Power Budget Calculation

The power budget for network hardware depends on the dynamic range, the difference between the sensitivity of the receiver and the output of the source into the fiber. You need some margin for system degradation over time or environment, so subtract that margin (as much as 3dB) to get the loss budget for the link.

Step 5. Data From Manufacturer's Specification for Active Components (Example: 100 Mb/s multimode digital link using a 850 nm LED source.)

Operating Wavelength (nm)	1300
Fiber Type	MM
Receiver Sensitivity (dBm@ required BER)	-31
Average Transmitter Output (dBm)	-16
Dynamic Range (dB)	15
Recommended Excess Margin (dB)	3

Step 6. Link Margin Calculation

Dynamic Range (dB) (above)	15	15
Cable Plant Link Loss (dB @ 1300 nm)	3.8 (Typical)	7.05 (TIA)
Link Loss Margin (dB)	11.2	7.95

In the past, as a general rule, the Link Loss Margin was expected be greater than approximately 3 dB to allow for link degradation over time or splicing for restoration. LEDs or lasers in the transmitter may age and lose power, connectors or splices may degrade or connectors may get dirty if opened for rerouting or testing. If cables are accidentally cut, excess margin will be needed to accommodate splices for restoration.

Today some systems, particularly high bit rate multimode LANs, have little margin due to the high bandwidth required. Some of these links require assuming fiber and connector loss to be extremely low to even accommodate the small power budget available. Under such conditions, you should assume lower values, especially for connector loss, and, of course, require installers to be extremely careful in installation to meet these needs.

Note: FOA offers a free App for smartphones and tablets to calculate loss budgets. Check the apple App Store for "FOA LossCalc."

Mixing Fibers
One problem that exists in designing and installing fiber optic networks is the possibility of mixing the types of fibers in a cable plant. This needs to be considered when doing loss and power budgets.

Singlemode: Mixing Singlemode Fiber Types
The different types of SM fiber are aimed at specific types of networks depending on the length of the links, transmission wavelength and use of DWDM and fiber amplifiers. Many of these fibers have different mode field diameters and that can create directional differences in losses when mated.

Generally speaking, mixing these fibers will not cause large variations in splice or connector loss, so it is not necessary to be overly concerned about mixing them. Thus, pigtails of one type of fiber can be used for termination of other types of fiber with minimal excess losses. However, when designing networks, long lengths of fibers should be of the same type to preserve the integrity of the system and ensure no problems with future upgrades to higher speed systems.

Multimode: Power Penalty For Mixing Multimode Fiber Types
Most premises installations now use 50/125μ (micron) fiber for Gigabit and
10 Gigabit Ethernet instead of the 62.5/125μ fiber used from the mid-1980s
to 2000. The laser-optimized 50/125μ fiber is a better solution for high speed
systems with VCSEL sources and is still compatible with virtually every
system that would also run on 62.5/125μ fiber.

A problem is the complications of mixing the two fibers in one installation,
a possibility in upgrades to older cabling systems. Mixing the two, as can
happen with installations that have both fibers and need both types of
patchcords, can induce severe losses in links. New TIA standards call for
color-coding cables and patch panels are designed to prevent mix-ups but not
all cabling follows these color codes.

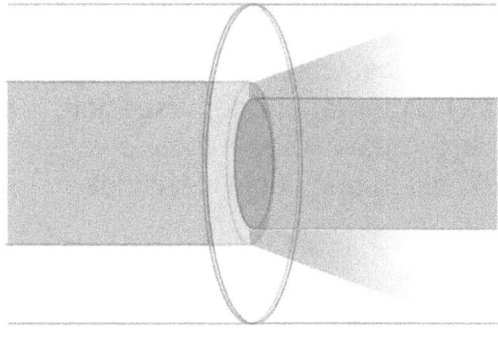

62.5/125 fiber   50/125 fiber

Fiber mismatch losses are directional. If you connect a 50/125μ fiber to a
62.5/125μ fiber, the smaller core of the 50/125μ fiber easily couples to the
62.5/125μ fiber and is very insensitive to offset and angular misalignment.
However, in the reverse direction, the larger core of 62.5/125μ fiber overfills
the core of the 50/125μ fiber, creating excess loss when light travels in that
direction.

The table below includes the penalties for mismatched multimode fibers.
Included are two obsolete fibers, 85/125μ and 100/140μ fibers. We include
100/140μ even though it is rare since we often get questions about old
systems using this fiber and the system owners are having difficulty finding
cables made of that fiber.

Loss Penalties For Mismatched Multimode Fibers

Receiving Fiber	Transmitting Fiber		
	62.5/125	85/125	100/140
50/125	0.9-1.6 dB	3.0-4.6 dB	4.7-9.0 dB
62.5/125	-	0.9 dB	2.1-4.1 dB
85/125	-	-	0.9-1.4 dB

There can be much variability in the power penalty, especially when if the system uses a 850 VCSEL source with its narrower modal fill in the fiber, rather than a LED used for slower systems and most test equipment. That is the reason for the range of values given in the table above – lower losses are for systems using VCSELs and the larger values are for LEDs.

The data is very important, however. The excess loss caused by coupling 50/125 fiber to 62.5/125 fiber enacts a penalty on power budget, especially when for used gigabit and above VCSEL source networks with their lower power margins. Mixing 50/125 and 62.5/125 fiber in one cable run is not recommended.

The best way to segregate 50/125 and 62.5/125 fiber in one installation is to color code the cables to keep them separate and/or use different connectors. Since the LC connector is becoming the de facto standard for gigabit and above, using LCs for 50/125 fiber is a bulletproof solution.

Mixing OM2/3/4 Fibers And The Effects On Bandwidth
OM2/3/4 fibers are all the same size but are manufactured to different standards that affect bandwidth. You should not see excess losses when connecting these fibers but it may affect total bandwidth. It is recommended that the patchcords be of a grade with the same or higher bandwidth than the installed cable plant.

Mixing Regular And Bend Insensitive Fibers
Most MM fiber is now of a type called "bend insensitive" (BI) that uses a special fiber structure to reflect light lost in bends back into the core. This type of fiber structure seems to not cause problems when SM fibers are mixed but may cause higher loss in MM fibers. The problem is the BI structure affects the mode power distribution in MM fiber, reflecting the light back into the higher order modes of the MM fiber. Connecting a BI fiber to a non-BI fiber may cause higher loss than mating like fibers and the opposite, lower loss, mating non-BI fibers to BI fibers. The differences are only a few tenths of a dB, but may be a problem in high speed MM systems with low loss margins.

You should try to avoid mixing BI and non-BI fibers in a network. Testing is another problem but somewhat different. Since it is hard to control mode power distribution in BI MM fibers, reference cables for loss testing are generally recommended to be non-BI fibers. Data shows the losses are directional and can cause problems measuring loss with OTDRs.

Comparing Loss Budgets To Test Results

A major use of loss budgets is estimating cable plant losses when properly installed. These estimates can be used as pass/fail criteria when testing the cable plant, but doing so requires using some judgment. Remember the loss budget is an estimate based on some specifications that you assume are appropriate for the cable plant being installed. Test results are subject to uncertainties also, caused by combinations of instrument and operator error. And it's possible that the test method's 0dB reference method and the loss budget's consideration of connector loss (see above) are not based on the same methodology.

Thus when comparing test results to loss budgets, use some judgment. If the loss budget is 2.15dB and the test result is 2.17, don't immediately consider the fiber tested to be a failure. The loss budget could be uncertain by several tenths of a dB as could the measurement.

If the measured loss exceeds the loss budget by a significant amount on a number of fibers, it's time to do some troubleshooting, starting with visual inspection and cleaning of the connectors, then retesting. Most installation and testing problems are traceable to dirty connectors, but if you find bad connectors, they may require replacement.

Chapter Exercises
- Calculate loss budgets for these networks, using both typical and worst-case estimates of connector loss:
- Data center: 100m links, LC connectors, 4 connections, both multimode fiber at 850nm (VCSEL) and singlemode fiber at 1300nm (laser)
- Metropolitan network: singlemode fiber, 15km, SC connectors – 5 connections, fusion splices – 7 splices
- FTTH network: singlemode fiber, 6km, 32 port split (17dB), 3 SC connections, 4 fusion splices.
- Long distance network: singlemode fiber, 65km, 13 fusion splices, SC connectors on each end

Chapter Quiz

1. A loss budget is the calculated loss of the cable plant while a power budget is the optical loss tolerable to a communications system.
    True
    False

2. Loss budgets are used to ensure _____.
    A. The network design will work with the chosen communications equipment
    B. Losses of components chosen are appropriate for the cable plant
    C. The cable plant tests have a comparison for pass/fail decisions
    D. All of the above.

3. When calculating the loss budget of a cable plant, you total the losses of all the _____ in the link.
    A. Fiber attenuation
    B. Connections
    C. Splices
    D. Passive devices
    E. All of the above

4. When calculating the loss budget, you should choose the component losses using _____.
    A. Loss values from industry standards that are always worst case
    B. Typical losses that are generally lower than standards
    C. Either typical or standard losses as long as it's documented in the design
    D. Lowest possible losses so the cable plant loss budget looks better

5. You calculate the contribution of the loss of the fiber to the loss budget by _____.
    A. Looking up the attenuation of the fiber on a manufacturer's data sheet
    B. Dividing the length of the fiber by the attenuation
    C. Multiplying the length of the fiber by the attenuation coefficient
    D. Choosing the best loss possible

6. When calculating the contribution of the fiber loss to the loss budget, you must consider the _____.
    A. Size of the fiber
    B. Type of cable
    C. Termination of the fiber
    D. Wavelength of the light in the fiber

7. Connector losses are calculated by adding up all the losses of the connectors, always _____.
   A. Including the connectors on each end of the cable plant
   B. Including the connectors on each end of the cable plant only if they are connected to a patchcord
   C. Excluding the connectors on each end of the cable plant
   D. Excluding the connectors on each end of the cable plant if the cable is connected directly to a transceiver

8. A premises cabling link 100 meters long uses multimode fiber (3.0 dB/km @ 850nm) and two connections in the middle as well as two connectors on the ends (0.50 dB/connector). The calculated loss budget would be

   _____.
   A. 1.30dB
   B. 2.30dB
   C. 3.30dB
   D. 5 dB

9. Recalculate the loss budget of the premises cabling link above (100m with 2 connections and connectors on each end) using TIA 568 worst case component losses (fiber at 3.5dB/km and connections at 0.75dB). Then the loss budget now becomes _____.
   A. 1.35dB
   B. 1.85dB
   C. 3.35dB
   D. 6.50dB

10. When comparing calculated loss budgets to test values of the installed cable plant in the field to determine whether an installation is acceptable, it's important to remember _____.
   A. The loss budget is an estimate
   B. The test results have some errors
   C. The operator must use judgment when the loss measured is close to the loss budget
   D. All of the above

Chapter 7

Optical Power Measurement

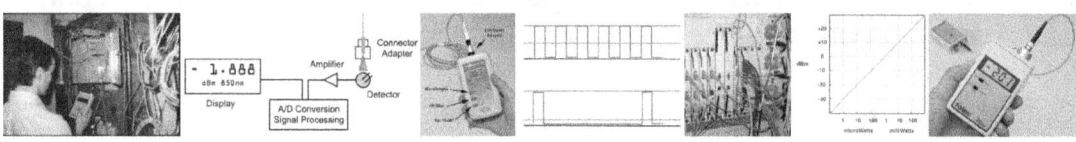

Objectives: From this chapter you should learn:
What optical power is
Why optical power is important in fiber optic measurements
How optical power is measured
How optical power meters are calibrated
What a dB is and why it is used in measurements
What causes errors in making optical power measurements

Optical Power

Practically every measurement in fiber optics requires measuring optical power. Measuring optical power in fiber optics is the equivalent of measuring voltage in electrical systems, where voltage is the basis of most electrical measurements.

Optical power in fiber optics refers to the energy in a source of light, generally an optical signal exiting a fiber. The source of that light may be the output of a transceiver coupled into a patchcord being tested with a power meter. It could also be a fiber optic test source being used to measure loss, where you make two measurements, the output of the test source and the output of the test source after passing through a fiber in the cable being tested. The cable loss is the difference between the two power measurements.

The output of a transmitter and the input to a receiver are "absolute" optical power measurements, that is you measure the actual value of the power referred to a standard. Loss is a "relative" power measurement, the difference between the power coupled into a component like a cable, splice or a connector and the power that is transmitted through it. This difference in power level before and after the component is what we call optical loss and defines the performance of a cable, connector, splice, or other component.

Optical Power Meters

The instrument designed to measure power is called an optical power meter or fiber optic power meter. However, other fiber optic instruments like the optical loss test sets (OLTS), optical time domain reflectometer (OTDR), optical continuous wave reflectometers (OCWR) used for return loss testing, fiber identifiers, optical spectrum analyzer (OSA) and practically every fiber optic measurement instrument measures optical power.

Optical power is generally measured in decibels (dB.) For *absolute* power measurements such as the output of an optical transmitter or the input to a receiver dB referenced to one milliwatt of optical power (dBm) is used. For loss measurements that are a *relative* measurement of power, dB is used. The use of dB as a measurement unit is explained below.

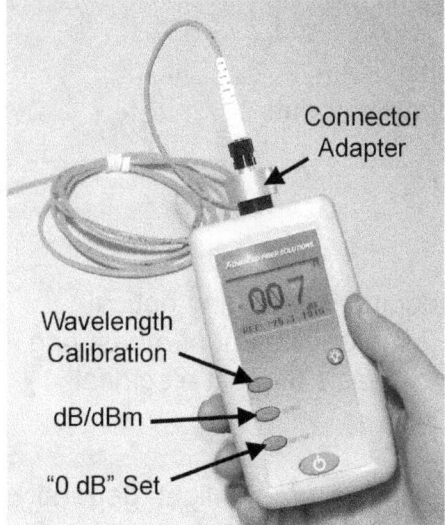

Typical fiber optic power meter showing controls and fiber interface

Optical power meters often cover a very broad dynamic range of power, typically over 1 million to 1 or 60 dB. Although most fiber optic power and loss measurements are made in the range of +10 dBm to -50 dBm, some power meters offer much wider dynamic ranges, especially if they are designed to measure the output of high power laser sources and fiber amplifiers. Here are some typical fiber optic systems and their ranges of power.

Typical power levels in fiber optic networks

Network	Wavelength (nm)	Power Range (dBm)
Telecom	1310, 1550	+10 to -40
Telecom WDM amplified	1500 to 1600	+20 to -40 per wavelength
CATV (HFC)	1550	+20 to -10
Datacom (LED)	850, 1300	-10 to -30
Datacom (VCSEL)	850	0 to -10
FTTH (GPON)	1310, 1490	+10 to -20

For testing analog CATV systems or fiber amplifiers, one needs special meters with extended high power ranges up to +20 dBm (100 mW). Although no fiber optic systems operate at very low power, below about -50 dBm, some lab meters offer ranges to -70 dBm or more, which can be useful in measuring optical return loss or spectral loss characteristics with a monochromator source.

Measuring Optical Power

Measuring optical power is straightforward. Most sources you will be measuring either have ports compatible to fiber optic connectors or are already attached to a fiber, either a cable or a pigtail. If it has an optical port but no fiber or cable, use a patchcord or reference cable to attach to the port. Select the proper connector adapter for the fiber optic power meter and attach it to the end of the cable.

Turn on the power meter and select the wavelength corresponding to the wavelength of the source. Choose the proper measurement range: dBm if making a measurement of absolute optical power, dB if measuring relative power as a reference for a loss measurement. Read and record the measurement with the relevant data, including source identification, wavelength, instrument identification and the reading of optical power.

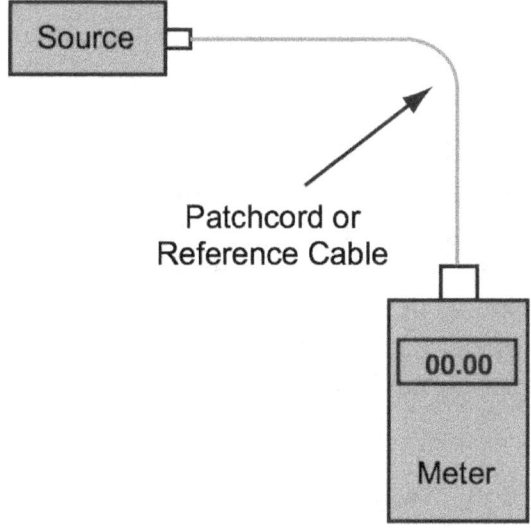

Measuring the optical power from a source

While the basic concept of measuring optical power is straightforward, there are some issues that require attention in order to get accurate data. The first issue is the choice of the test cable, which can be a regular patchcord or reference cable. Either may be used but it is important that the cables be in good condition and connectors are cleaned.

If testing a source by attaching a test cable of your choice, you need to choose a cable with the proper fiber, particularly with multimode systems using LED sources. The fiber size and numerical aperture will affect the power coupled from the LED to the test cable. Testing a LED source with 62.5/125 micron fiber which has a numerical aperture of 0.26 (OM1 fiber) will show much higher power, from 1 to 4 dB depending on the output pattern of the LED than when testing with 50/125 micron fiber with a numerical aperture of 0.20. If the source is being tested for use in a particular cable plant, it should be tested with the type of fiber being used in the cable system to get an accurate measurement of power when used there.

Testing VCSEL sources, the most common multimode source today because of its high data rate capability, should be done with only 50/125 micron fiber as that is the type of fiber that use VCSEL sources. Singlemode lasers should be tested with singlemode fiber test cables.

When testing power, as with all fiber optic tests, it is mandatory to clean all connectors to prevent attenuation of the light by dirt on the connectors. In addition, one should be careful to not put stress on the cable connecting the source to the meter. Stress on this cable, especially near the connectors, can cause loss in the cable that will attenuate the power being measured and affect measurement accuracy.

Understanding Decibels (dB)

Optical power measurements are expressed in decibels (dB,) the standard measurement unit of power and loss in optical fiber measurements. Optical loss is measured in "dB" which is a relative measurement, the difference between two measurements. Absolute optical power is measured in "dBm" which is dB referenced to 1 milliwatt of power. Loss is a negative number (like -3.2 dB) as are many power measurements. Measurements in dB can sometimes be confusing.

In the early days of fiber optics, source output power was usually measured in milliwatts, a linear scale, as was done in classical optics. But loss was measured in dB or decibels, a logarithmic scale. Over the years, all measurements migrated to dB for convenience causing much confusion. Loss measurements are measured in dB since dB is a ratio of two power levels, one of which is considered the reference value. Absolute power levels are measured in dBm where the reference value is 1 milliwatt (mW.)

The dB is a logarithmic scale where each 10 dB represents a ratio of 10 times. It can often be confusing, so let's look at it in more detail.

The actual equation used to calculate dB is:

dB = 10 $\log_{10}$ (measured power / reference power).

The relationship of power in dB to milliwatts is logarithmic, based on log 10.

dB = 10 $\log_{10}$ (measured power / reference power) *for relative power, e.g. loss*

and

dBm = 10 $\log_{10}$ (measured power / 1 mW) *for absolute power like transmitter output*

The graph below shows the relationship of dBm to watts *for absolute power*

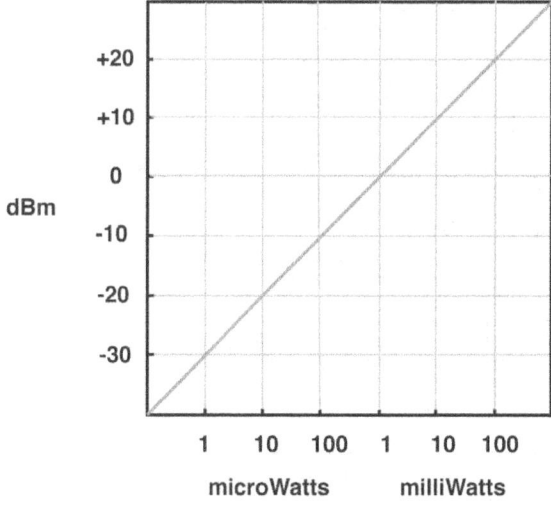

Relationship of dBm to watts

This graph of dBm vs. Watts helps understand how dBm and power in milliwatts (mW) are related when measuring absolute power. Note the following:
- 0 dBm on the vertical scale corresponds to 1 mW
- +10 dBm is 10 mW or ten times more than 1 mW
- -10 dBm is 100 microwatts that is 1/10th of a mW or 0.1 mW
- +20 dBm is 100 mW
- -20 dBm is 1/100 of a mW, 0.01 mW or 10 microwatts.

When measuring relative power for loss, we are measuring on a log scale but the actual measurement is a ratio of power as you see in the equation that defines dB.

$dB = 10 \log_{10}$ (measured power / reference power).

The table below shows the ratio of power for dB differences in power:

dB (gain)	Power ratio	dB (loss)	Power ratio
0	1.000	0	1.000
0.1	1.023	-0.1	0.977
0.2	1.047	-0.2	0.955
0.3	1.072	-0.3	0.933
0.4	1.096	-0.4	0.912
0.5	1.122	-0.5	0.891
0.6	1.148	-0.6	0.871
0.7	1.175	-0.7	0.851
0.8	1.202	-0.8	0.832
0.9	1.230	-0.9	0.813
1	1.259	-1	0.794
2	1.585	-2	0.631
3	1.995	-3	0.501
4	2.512	-4	0.398
5	3.162	-5	0.316
6	3.981	-6	0.251
7	5.012	-7	0.200
8	6.310	-8	0.158
9	7.943	-9	0.126
10	10	-10	0.1
20	100	-20	0.01
30	1000	-30	0.001
40	10000	-40	0.0001
50	100000	-50	0.00001
60	1000000	-60	0.000001

Compare the positive and negative dB across the rows. The ratio of the positive dB is the inverse of the negative dB, e.g. +10dB is a ratio of 10 times and -10 dB is a ratio of 1/10 or 0.1. Thus 10 dB is a ratio of 10 times: +10 dB means the power measured is 10 times greater than the reference power and

-10 dB is one-tenth as much. Some of the numbers are easy to remember and may be useful. For example, +3 dB is a factor of two in power and -3 dB is a factor of one-half.

When the two optical powers compared are equal, dB = 0, a convenient value that is easily remembered. If the measured power is higher than the reference power, dB will be a positive number, but if it is lower than the reference power, it will be negative. Thus measurements of loss are expressed as negative numbers.

Measurements of optical power such as the output of a transmitter or input to a receiver are expressed in units of dBm. The "m" in dBm refers to a reference power of 1 milliwatt. Thus a source with a power level of 0 dBm has a power of 1 milliwatt. Likewise, -10 dBm is 0.1 milliwatt and +10 dBm is 10 milliwatts.

To measure loss in a fiber optic system, we make two measurements of power, a reference measurement before the component we are testing and a loss measurement after the light passes through the component. Since we are measuring loss, the measured power will be less than the reference power, so the ratio of measured power to reference power is less than 1 and the log is negative, making dB a negative number when measuring loss. That can be confusing on some OLTS that are not calibrated as power meters. Some manufacturers have the display read positive dB for loss, ignoring industry convention and confusing many techs.

When we set the reference value for loss testing, the meter should read "0 dB" because the reference value we set and the value the meter is measuring is the same. Then when we measure loss, the power measured is less, so the meter should read a more negative number, "- 3.0 dB" for example, if the tested power is half the reference value.

Although fiber optic power meters measure a negative number for loss, on some optical loss test sets (OLTS) the loss is expressed as a positive number, so the loss is displayed as 3.0 dB when a fiber optic power meter would read - 3.0 dB.

Absolute Power Calibration And Measurement Uncertainty

FO power meters should be calibrated to transfer standards provided by national standards laboratories like the US National Institute of Standards and Technology (NIST). These national laboratories have primary standards used for calibration of optical power at the standard fiber optic wavelengths, 850,

1300 and 1550 nm. These standards are extremely stable and continuously monitored.

Standards labs create transfer standards consisting of a calibrated laboratory power meter and three laser sources at the proper wavelength that are shipped around to test equipment manufacturers. The manufacturer calibrates their working standards to these transfer standards and then uses their working standards to calibrate instruments they sell to customers.

A properly calibrated fiber optic power meter will have a typical measurement uncertainty of ±5% or about ±0.2 dB when measuring absolute power. This uncertainty is caused by the limitations of the calibration process.

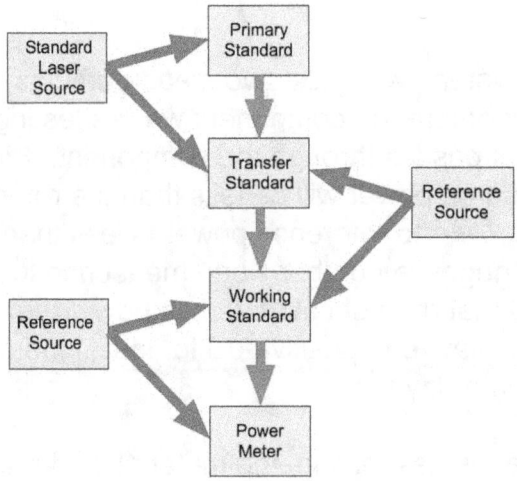

Calibration of optical power meters

When you buy a fiber optic power meter, it is 3 steps removed from the primary standard: meter to working standard, working standard to transfer standard, transfer standard to primary standard. Each step of that process has some inherent error, in the range of 1%. When you make a measurement of optical power with that meter, you add a fourth step with somewhat more inherent error so the total measurement uncertainty for absolute optical power is about +/- 5%.

Calibration is generally done at several power levels to ensure the meter response is linear, that is the meter reads properly over a range of optical power. Rather than change the output of the source, attenuators are used to reduce power by known levels to check calibration. The calibration process is often automated and the calibration data is stored in the memory of the power meter.

Sources of errors when measuring power are the variability of coupling efficiency of the detector and connector adapter, reflections off the shiny polished surfaces of connectors, source wavelengths (since the detectors are wavelength sensitive), precision of the attenuators used, nonlinearities in the analog electronic signal conditioning circuitry or the analog to digital converter of the FO power meter and detector noise at very low signal levels. These factors will be covered in more detail below and in Chapter 9 on Metrology.

Since most of these factors affect all power meters, regardless of their sophistication, expensive laboratory meters are hardly more accurate that the most inexpensive handheld portable units.

Meters should be recalibrated according to manufacturer's directions by labs with traceable calibration systems. Many standards call for meter calibration to ensure accurate measurements.

Relative Measurement Errors

When measuring loss, meters are not measuring absolute power in dBm but relative power in dB. Relative power is not affected as much by calibration but is affected by the linearity of the meter's performance.

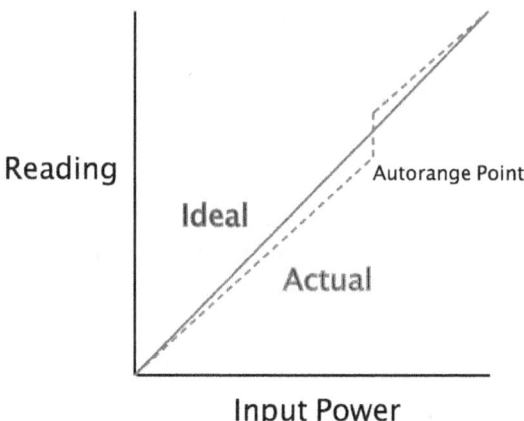

Effect of autoranging on optical power meter accuracy

The linearity of the meter can be affected by two issues - the slope of the calibration curve and the offset where the meter autoranges. The linearity of the slope is important since it determines if the dB reading on the meter is scaled correctly to read the input power correctly.

Most meters read power over a large dynamic range, 1,000,000 to 1 (60dB) is not uncommon, and that is beyond the range of the electronics in the meter. The analog amplifiers which condition the output of the detector have a linear output and dB is calculated by the microprocessor controller. The range of the amplifiers is less than the specified range of the meter. To get a larger range, the meters have a range change that is controlled by the microprocessor that changes the gain in the amplifier attached to the detector.

If a meter is being used to measure loss and the source output is near the autorange point, the reading can be affected by the nonlinearity caused by autoranging. Calibration of the meter should include looking at the linearity around the autoranging points to ensure minimal nonlinearity.

Understanding FO Power Meter Measurement Uncertainty

Much attention has been paid to developing transfer standards for fiber optic power measurements. The US NIST in Boulder, Colorado and standards organizations of most other countries have worked to provide good standards to work from. We can now assure traceability for our calibrations, but even so the errors involved in making measurements are not ignorable. Even when fiber optic power meters are calibrated within specifications, the uncertainty of a measurement may be as great as ±5% (about ±0.2 dB) compared to primary standards. Understanding power meter errors and their probable causes will insure a realistic viewpoint on fiber optic power measurements.

The first source of error is optical coupling. Light from the fiber is expanding in a cone. It is important that the detector to fiber geometry be such that all the light from the fiber hits the detector, otherwise the measurement will be lower than the actual value. But every time light passes through a glass to air interface, such as the window on the detector, a small amount of the light is reflected and lost. Even the cleanliness of the optical surfaces involved can cause absorption and scattering. The sum total of these potential errors will be dependent on the connector type, wavelength, fiber size and NA.

There are errors associated with the wavelength calibration of the meter. Semiconductor detectors used in fiber optic instruments (and systems too) have a sensitivity that is wavelength dependent. Since the actual source wavelength is rarely known, there is an error associated with the spectral sensitivity of the detector. By industry convention, the three cardinal wavelengths (850, 1300 and 1550 nm) are used for all power measurements, not the exact source wavelength. However, variations in the sensitivity of detectors used in meters may cause errors in measurement, although those errors should be relatively small.

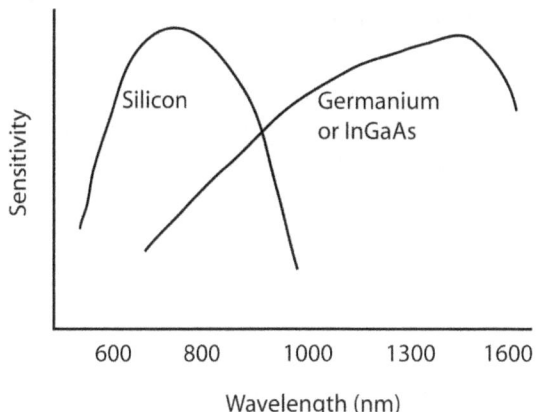

Sensitivity of detectors used in fiber optic power meters

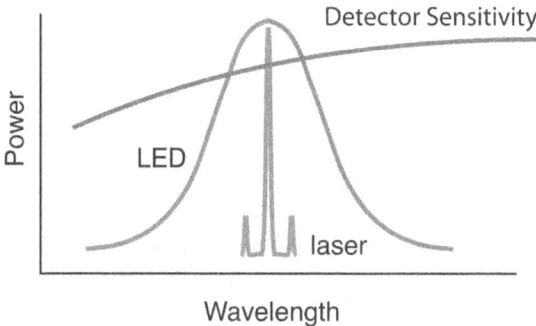

Spectral width of sources and detector sensitivity

Measurement variations may also occur due to the spectral characteristics of sources. LEDs have broad spectral widths while the power in lasers is concentrated in a very narrow spectral range. The measured power of lasers can vary with detector sensitivity while variations in the shape of the spectral output of LEDs may make measurements vary. These are factors that are hard to control even in a lab environment. In the field, they simply add more uncertainty to the measurement.

Another source of error exists for high and low level measurements. At high levels, the optical power may overload and saturate the detector, causing the measurement to be in error. Special meters need to be used to measure the outputs of higher power lasers like DFB lasers or fiber amplifiers. At low levels, the inherent detector noise adds to the signal and becomes an error. If the signal is 10 dB above the noise floor (10 times the noise), the offset error is 10% or 0.4 dB.

Instrument Resolution vs. Measurement Uncertainty

Considering the uncertainty of most fiber optic measurements, instrument manufacturers have provided power and loss meters with a measurement resolution that is usually much greater than needed. As noted above, the uncertainty of absolute optical power measurements is about ±5% (about ±0.2 dB), loss measurements are likely to have uncertainties of ±0.2 to 0.5 dB or more, and optical return loss measurements have a greater than ±1 dB uncertainty.

Fiber optic power meters generally have a display resolution of 0.01 dB. That is considerably more than the uncertainty of most measurements. It is common and preferable, however, to have the resolution of an instrument be ten times the typical measurement uncertainty of the instrument to allow for making small measurements more easily.

If you are trying to measure small changes in loss, like stress effects on a cable, or testing a cable plant with a total loss of only 2 dB, more instrument resolution is helpful. Within the laboratory, a resolution of 0.01 dB can be extremely useful, since one often measures the loss of connectors or splices that are under 0.10 dB or changes in loss under environmental stress that are under 0.10 dB. Stability of test sources and physical stress on cables limits measurement uncertainty to about 0.02 to 0.05 dB per day, but 0.01 dB resolution can be helpful in determining small changes in component performance.

Field measurements have higher uncertainty because more components are measured at once and losses are higher. Practically, measurements are better when the instrument resolution is rounded off to 0.1dB. Readings will be more likely to be indicative of the measurement uncertainty.

Chapter Exercises
- Connect a power meter to a fiber optic test source with a patchcord. Turn on the source and let it warm up until it stabilizes. Set the meter to the wavelength of the source and measure the power. Change the wavelength to all the other wavelengths calibrated on the meter. What happens? Why?
- If your power meter has a linear scale, measure the output of a source and change between dB and watts. Do the readings correspond to the table in this chapter?
- If you have access to more than one power meter, use each of them to measure the output of a source. Do they all read the same? Why or why not?

Chapter Quiz

1. Optical power is the equivalent to _____ in electrical systems.
    A. Voltage
    B. Current
    C. Resistance
    D. Impedance

2. Optical power is physically defined in units of watts but is usually measured in a logarithmic scale and expressed in _____.
    A. Milliwatts
    B. Microwatts
    C. dB
    D. Ohms

3. A measurement in dB is a relative power measurement, for example when testing _____.
    A. Transmitter power
    B. Receiver power
    C. Cable plant loss
    D. Bandwidth

4. A measurement in dBm is an absolute power measurement, for example when measuring _____.
    A. Transmitter or receiver power
    B. Connector loss
    C. Cable plant loss
    D. Bandwidth

5. The "m" in dBm means the optical power is _____.
    A. Measured by a "meter"
    B. Referenced to another "measurement"
    C. Referenced to "1 milliwatt"
    D. A "mandatory" measurement

6. The difference between two measurements in dBm is expressed in dB, for example in measuring loss.
    True
    False

7. A 3 dB loss in the cable plant means the optical power has changed by a factor of _____, while 10 dB is a factor of _____.
    A. 2, 10
    B. 2, 100
    C. 20, 1000
    D. 10, 100

8. Fiber Optic power meters are calibrated at different wavelengths because the sensitivity of their detectors varies with wavelength.
    True
    False

9. OTDRs measure optical power on the vertical scale and _____ on the horizontal scale of a fiber trace.
    A. Pulse width
    B. Distance
    C. Resolution
    D. Reflectance

10. The attenuation coefficient of a fiber as measured by an OTDR is calculated in _____.
    A. dB
    B. dBm
    C. dB/km
    D. dBx

Chapter 8

Insertion Loss Measurement

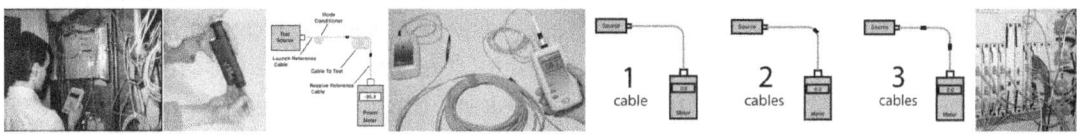

Objectives: From this chapter you should learn:
What is optical loss
Testing methods and standards
Choosing appropriate test equipment
Options for setting the 0 dB reference and testing various connector types
Test conditions for different fiber types
Sources of error
Estimating test results with loss budgets
Recording data for documentation

What is Insertion Loss?

Optical loss is a term used in many contexts in fiber optics. It simply means a reduction in optical power, for example the loss caused by a component or an entire cable. The component could be a length of fiber, a splice, a connection made between two connectors or a passive component like an attenuator, splitter or switch. The cable could be a patchcord or an installed cable plant.

The primary test for these is an insertion loss test, test that uses a test source and optical power meter to measure the difference in power when the component is inserted in the test setup. For a cable plant, the insertion loss test uses a the test source and power meter to simulate the transmitter and receiver of a communications link. Variations of this test are used for practically every loss test in fiber optics.

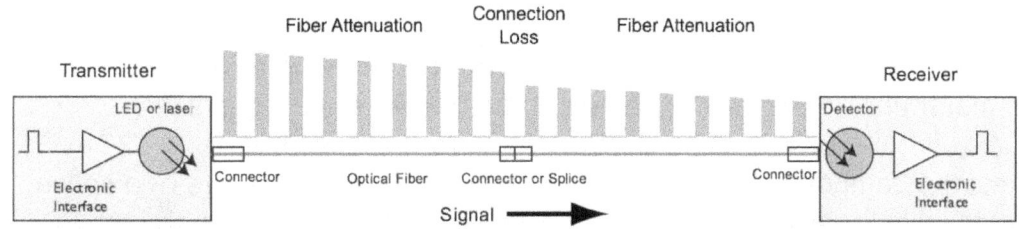

Optical loss in a fiber optic link

The goal of an insertion loss test is to simulate link operating conditions by using a test source to launch power into the cable to test and a power meter to measure the loss at the other end. That requires creating launch conditions from the test source that are similar to transmitter sources and using the power meter to measure the power from the source before and after the component or cable is inserted in the test setup. How this is done is the subject of many standards in fiber optics that ensure the test results from laboratories, manufacturers, installers or users are comparable.

Insertion Loss Test

Why is this test called an "insertion loss" test? The name comes from the fact that one performs the test by inserting the component under test between a test source and a power meter. The most obvious version of this test is the method used by manufacturers to evaluate connector or splice loss as shown in the diagram below. It can also be used to measure the attenuation of optical fiber.

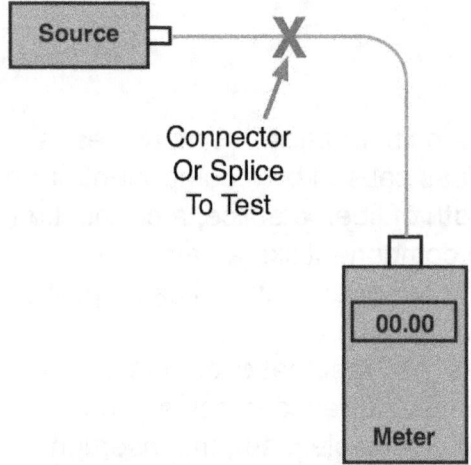

Insertion loss test diagram - component loss

This is the insertion loss test used by manufacturers for evaluating the performance of connectors or splices. This test connects the test source to a power meter over an optical fiber. Sometimes this test is done with bare optical fiber with the connection to the power meter using a bare fiber adapter. The meter and source are turned on and a "0 dB" reference is set.

To test the components, the fiber is cut and a pair of connectors or a splice is inserted in the fiber and the change in power measured. The change in power indicates the loss generated by the insertion of the components. When evaluating a new component, manufacturers may do hundreds or thousands of these tests to determine the average performance of the component that they specify in their product datasheets.

For tests involving multimode fiber, the test results are highly dependent on the test setup, particularly the modal distribution launched into the fiber used in the test. Test sources can have significant variations in modal distribution so it is important to include some form of modal conditioning in the test fiber. Modal conditioning and other issues that affect all multimode tests will be covered in a section below on test conditions.

Note: *It's common in fiber optics to talk about the loss of a connector, but a connector loss is measured when mated to another connector. In fact, a single connector has no loss because a connector is defined as a component that allows making connections between two fibers or connects one fiber to an active device like a transmitter or receiver. The correct term is "connection loss" because it is the loss of a mated pair of connectors. In practice, we assign a loss to a connector by testing it against a reference connector, as you will see when we discuss single-ended insertion loss testing below.*

Cutback Test For Bare Fiber

There is another variety of insertion loss testing used to test bare optical fibers. This test is called a "cutback test" because of the way it is performed. In a cutback test a long length of fiber is connected between the source and the power meter using special temporary connectors called bare fiber adapters. The power is measured by the meter and recorded. The fiber is then cut back to a point near the source and the power measured again. The change in power is the loss of the optical fiber. Knowing the length of the fiber in kilometers allows calculating the attenuation coefficient of the fiber:

dB loss / length = attenuation coefficient in dB/km

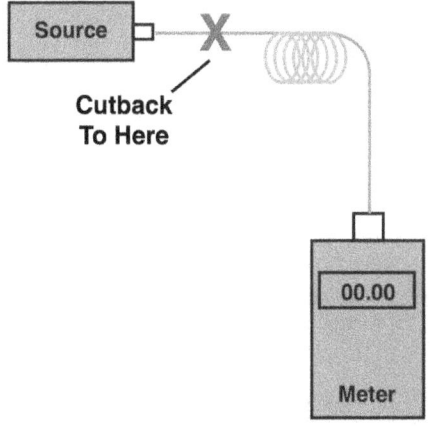

Cutback test on bare fiber

A variation of this test can be performed with a pigtail (a short cable with a connector on one end and bare fiber on the other end) attached to the fiber under test with a mechanical splice for a temporary connection. In that case the test results are less accurate due to the unknown loss in the splice or connection.

As with insertion loss testing, cutback tests involving multimode fiber are highly dependent on the optical test setup, particularly the modal distribution launched into the fiber used in the test. Modal conditioning and other aspects of the test setup affect all multimode tests so it will be covered in a section below on test conditions.

Testing Cables And Cable Plants

The most common tests involve testing fiber optic cable plants, patchcords and other components that have connectors on each end. These are tested with a test source and power meter with reference test cables to connect to the component under test. There are two versions of this test, a double ended test and a single ended test.

The double-ended test is the standard test for installed cable plants that allows testing the entire cable plant including the connectors on each end. It is a simulation of the loss an actual transmission system will see when connected to the cable with patchcords.

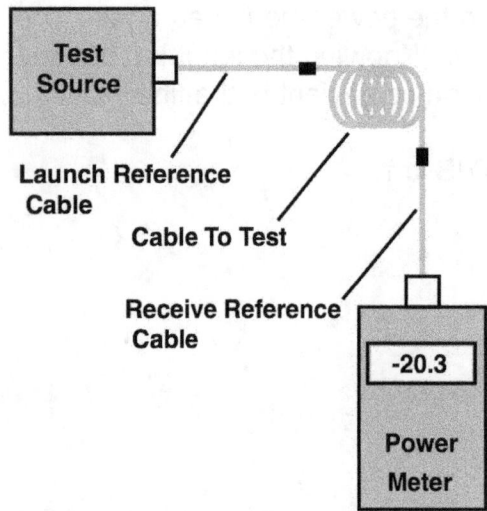

Insertion loss test diagram – double ended test

The source has a launch reference cable because it launches power into the component under test. The power meter has a receive reference cable because it receives the power after the component has been inserted.

The "0 dB" reference measurement is made with the test source and power meter with their reference cables using one of three different but acceptable methods. The insertion loss test then measures the loss generated by adding the cable under test. The value measured is called the loss of that particular fiber optic cable.

That test is called a "double ended" loss test because it uses reference test cables at the source, the launch cable, and the meter, the receive cable, to measure the loss of the connectors on both ends of the cable under test. The cable under test is sometimes called the "permanent link," a reference to the name of the installed cable in UTP copper cable testing.

Testing Patchcords

One can also do a variation of the insertion loss test with only a launch cable, called a "single ended" test. A single-ended test only tests the connector on the source end of the cable as it is mated to the reference cable connector. This test is often used to test patchcords since it allows the measurement of the connectors on each end separately, a more rigorous method of testing these short cables.

The single ended test looks like the double ended test except there is no receive reference cable. The cable under test is mated to the connector on the launch cable and the power meter. The "0 dB" reference measurement is made with the test source, a launch reference cable and the power meter, exactly like the one cable reference for double ended insertion loss tests.

The output of the launch cable is the 0 dB loss reference, so for short cables like patchcords, the measurement is basically the loss of the connection between the launch cable and the test cable. The test cable can be reversed and the connector on the other end tested separately. This test is used on patchcords because it allows testing each connector individually, ensuring that both connectors meet specifications. If the cable were tested with the double ended test, one would measure a total loss of both connectors and could not identify if one connector were out of specification.

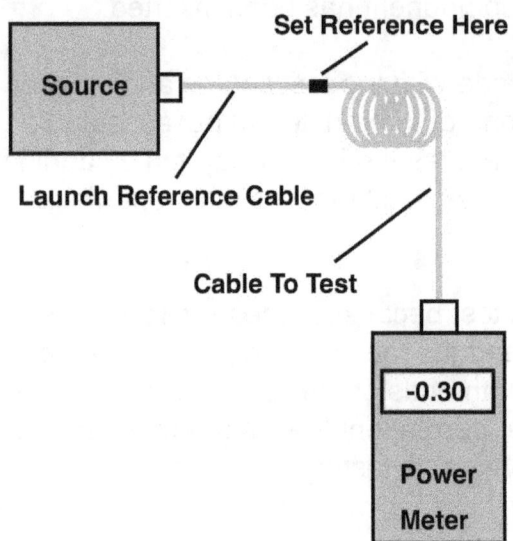

Insertion loss test diagram – single ended test

Standards For Insertion Loss Testing

The majority of fiber optic standards are related to testing. Many are specific to components and are used by manufacturers to determine the specifications that they cover on their data sheets. Some standards are used to qualify components for specific applications, e.g. use in adverse environments or under stress. Only a few actually relate to basic fiber optic testing such as those used during installations. The appropriate procedures are well covered by manufacturers' directions, negating the need for most users to refer to the actual standards. A listing of international standards relating to fiber optics is included in Chapter 3.

However, many installation contracts will refer to these standards so it is imperative to understand what they cover to ensure meeting the conditions of the contract. In this chapter the content and conditions of these standards will be described to ensure the reader understands what these standards require for testing.

Testing the complete cable plant is covered in many standard test procedures, e.g. FOA Standards FOA-1, FOA-2 and FOA-7, TIA OFSTP-14 for multimode fiber or OFSTP-7 for singlemode. Both use the same procedures as ISO/IEC 61280-4-1, ISO/IEC 14763 and other international standards.

Early versions of the TIA and ISO/IEC standards were straightforward directions for making tests, but recent updates have included more about test conditions than actual testing procedures. For example, you must go to the back of the latest TIA OFSTP-14 or ISO/IEC 61280 standards on testing multimode fiber for premises applications to find information on how to actually conduct the test. The majority of the standard relates to mode power control in current generation multimode fiber (OM3 and OM4) that is an important aspect of testing but confusing to readers when it precedes how the test is actually conducted.

To help users understand these different test standards, FOA has created a series of "1 Page Standards" that provide concise summaries of all the basic tests for optical loss. These standards are compatible with the processes included in the standards and offer advice on how to conform to the standards.

The standards covering insertion loss offer many options, for example 3 different ways to set a "0 dB" reference using one, two or three reference cables, but often do not provide guidance on where and why these options should be used. Standards for multimode fiber testing always cover the problems of controlling mode power distribution, a major source of measurement uncertainty. This chapter will cover these issues in detail.

Choosing Appropriate Test Equipment

In order to perform an insertion loss test, it is necessary to have appropriate equipment for the cables or cable plant under test. Whether one uses a test source and power meter or optical loss test set (OLTS), the equipment must be compatible with the test requirements.

The fiber optic power meter used for insertion loss testing should be calibrated at the wavelength of the test source being used. The meter should have a connector adapter compatible with the connectors on the cable plant being tested. Having a special "dB" range that will allow setting a "0 dB" loss reference power level will simplify testing.

An OLTS will offer both source and power meter for loss testing at the appropriate wavelengths. The requirements for the source for modal conditioning (see below) and compatibility of the instrument to the connectors on the cable plant being tested may have to be accommodated by the reference test cables.

Test Source Wavelengths
For multimode fiber, the test source should be a LED at 850 nm that is the wavelength used for virtually all multimode communications systems.

There is an option for testing at 1300 nm with a LED also, but there are few systems today that operate at that wavelength so testing at that wavelength is generally unnecessary. You will find some references to using 1300 nm testing to find stress on the cable, since multimode fiber is much more sensitive to bending losses at 1300 nm. But even that reason is no longer relevant in most cases because most multimode fiber is of the bend-insensitive (BI) type. Finding stress areas in cables can be tricky anyway, but if the cable is long enough, an OTDR that offers both 850 and 1300 nm testing is a better instrument for finding the location of the loss.

While most 850 nm multimode systems operating over OM3 or OM4 fiber today use VCSEL sources (vertical cavity surface-emitting lasers), these sources are not recommended for use as test sources due to the unpredictable mode power distribution of individual devices. Instead, standards now call for sources with mode power distribution or conditioning the output of the source and its launch cable to approximate an ideal VCSEL mode condition. This will be covered below in reference cables and again in the section on mode conditioning for multimode fiber testing.

Singlemode fiber is tested with laser sources, similar to the devices that will be used in the communications systems which operate over the fiber. Singlemode fiber will be tested with 1310 nm and/or 1550 nm lasers depending on the cable plant to be tested. Short links, from hundreds of meters in a data center or building up to 20 or 30 km in metro networks, are always tested at 1310 nm and often at 1550 nm if wavelength division multiplexing (WDM) or passive optical networks (PONs) like FTTH are planned for use on the link. Long links are tested at 1550 nm to match the wavelengths of networks using them.

There is also a reason to test short singlemode links at 1550 nm, to find stress loss in a cable plant caused by installation. But as mentioned above, this is not relevant if bend-insensitive fiber is being used in the link, common today in high fiber density cables. Also testing at 1550 nm for finding stress is more relevant to ODTR testing where the cause of the stress loss can be determined.

Directional Loss Measurements
Some past standards have required bidirectional insertion loss tests so some OLTS offer bidirectional testing modes. There are small directional differences in the loss of many joints. These may be caused by differences in the fiber designs or manufacturing variations.

Within singlemode fibers, there are regular SM fiber, large MFD fibers, dispersion shifted fibers and bend-insensitive fibers, plus manufacturing tolerances that can cause fiber differences. Some of these fibers have very different fiber index profiles. Mixing these fiber types often occurs because of the standard practice of splicing regular singlemode fiber pigtails on cables with fibers of any type.

Multimode fiber has manufacturing variations in core size among each fiber type, variations in index profiles of various bandwidth grades and bend-insensitive fiber variations.

These directional differences in loss are small but and very hard to measure. They require very careful lab procedures to isolate other variables, not something that can be done with field OTDR or insertion loss testing. Bidirectional testing, therefore, offers little benefit.

Reference Test Cables
Just as important as the choice of test equipment is the choice of reference test cables for the launch cable and receive cable. The basic requirements for test cables is that they be about 1 to 2 meters long, match the size of fiber in the cable plant under test and have connectors compatible to the connectors on the cable plant.

Multimode graded index fiber in test cables should be 62.5/125 for OM1 cable plants or 50/125 for OM2, OM3, OM4 or OM5 fiber cable plants. There are no significant differences in types of 50/125 fiber so any type of this size fiber can be used to test any other type. Today most of the multimode graded index fiber is bend insensitive (BI) fiber. Many standards recommend not using BI fiber for reference test cables even if testing BI fiber cables, but this may not be possible. We'll discuss BI fiber in the section on modal conditioning for multimode fiber.

When testing step-index multimode cable plants using plastic optical fiber (POF) or plastic coated silica fiber (PCS), one must always choose a matching fiber for reference cables.

The connectors on the test cables should be PC polished (physical contact) and must be of very high quality, determined by having low loss when tested against each other in the single ended test mode. One should use low loss patchcords, typically under 0.3 dB, so the test results will be consistent. With use, the connectors will wear, even when cleaned frequently. If the patchcords exceed 0.5 dB over time, they should be replaced. Careful repolishing with diamond polishing film by an experienced tech may bring the loss back down. Or the cables should be replaced.

The new TIA OFSTP-14 and ISO/IEC 61280-4-1 standards for multimode call for 2 meter long "reference quality" test cables with connector loss of 0.1 dB. The standard explains that this may require special cables with selected fibers and selected connectors as well as higher quality mating adapters such as those used for singlemode fiber with ceramic mating adapters. For singlemode fiber the loss specified is 0.2 dB.

There are very few sources of "reference quality" test cables. Multimode cables with losses of 0.1 dB will degrade with use, showing the effects of the many mating cycles required when testing in the field. Using high quality cables with relatively loss is the practical solution. Cables with loss of 0.2 up to 0.5 dB maximum are generally adequate for testing multimode fiber.

The launch reference cable combines with the test source to create the modal conditions used for testing. If the source modal conditions are not specified, and they rarely are, the modal conditions can be modified by the user using special patchcords or modal conditioning on the cable itself. This is discussed in the section on modal conditioning below. Do not ignore modal conditioning in multimode testing. When coupled into multimode fiber, LEDs typically have higher modes than are specified by test standards. Mode conditioning will remove those higher modes and tests will be more consistent and will generally show lower loss.

Singlemode reference cables should also be high quality cables. The connectors must also be compatible to the type of connector on the cable plant being tested, e.g. PC (physical contact, color coded blue) or APC (angled physical contact, color coded green.) For singlemode fiber "reference quality" test cables, the loss specified is 0.2 dB. High quality singlemode patchcords should be 0.3 dB or less so this limit is not unreasonable.

Singlemode does not have mode conditioning standards like multimode fiber, but short launch patchcords connected to a typical laser source may support 2 modes for short distances and that can affect measurement results. To ensure the launch is singlemode, a mode filter made with a small loop of fiber 40 to 60 mm diameter (1.6 to 2.4 inches) in the launch cable will ensure singlemode launching at the end of the cable.

Testing The Installed Cable Plant – Double Ended Testing

The most common test performed in fiber optics is the basic loss test for a connectorized cable or the installed cable plant. This is the test described above as the "double-ended loss test" and the diagram looks like this, with added notations that will be discussed below:

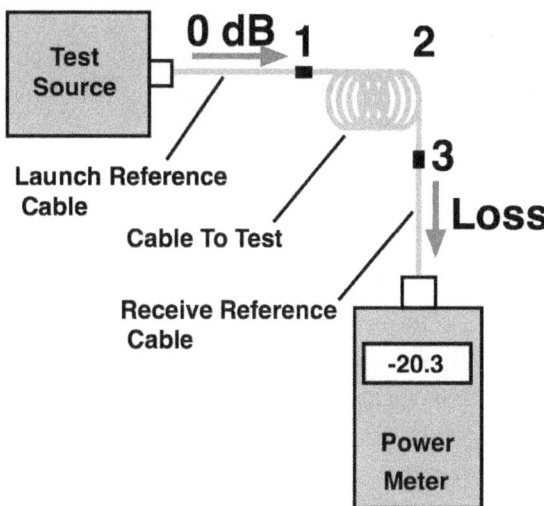

Cable plant insertion loss test diagram – double-ended test

As described above, the test involves inserting the cable to be tested between a source with a launch cable and a meter with a receive cable. The output of the source (noted as "0dB," is attenuated by the connection to the cable under test at point "1," further attenuated by the cable to test and finally by the connection "2" to the receive cable attached to the power meter.

What the power meter measures is (with reference to the numbers in the drawing above):

    Loss of launch cable connection (1)
 + Loss of components in the cable plant (2)
 <u>+ Loss of receive cable connection (3)</u>
= Total loss in dB

The components in the cable plant include all the losses caused by the attenuation of the optical fiber, any intermediate splices or connections, and any passive components like splitters used in FTTH (fiber to the home) or OLAN (optical LAN) PON networks.

Setting A "0 dB" Reference
How the meter translates this measurement into a loss figure in dB depends on how the reference power for "0 dB" is set. There are three methods used for setting the 0 dB reference because there are different test conditions that must be accommodated. These three 0 dB reference options look like this:

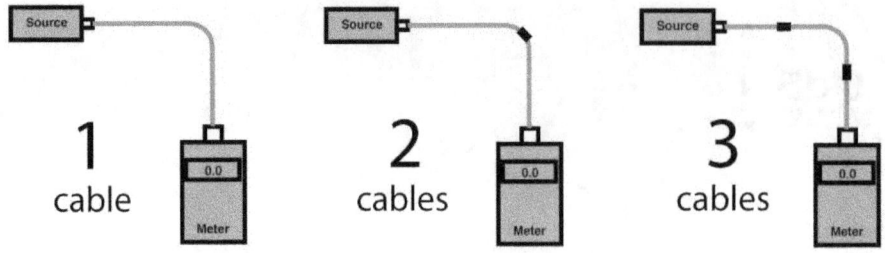

Three ways to set the "0 dB" reference for loss testing.

Many fiber optic test standards allow for all three methods but some only mention the 1- and 3-cable methods. The reason for the existence of three methods is the compatibility of test equipment to the cable plant; whether the test equipment has connector interfaces that allow direct connection to the cable under test.

The options for use of these three methods are:
- If the test equipment has connectors compatible with the cable plant, a one-cable method can be used.
- If the test equipment does not have connectors compatible with the cable plant, a two- or three-cable method must be used.
- If the test equipment does not have connectors compatible with the cable plant and the connectors are the "plug and jack" or "male, female" type, a three-cable method must be used.

Note: After setting the "0 dB" reference, it is important to not disconnect the reference launch cable from the source. The connection between the source and cable can change if it is disconnected and reconnected invalidating the "0 dB" reference set. The cables can be disconnected at the meter end because the meter has a large area detector that makes connections consistent.

Note: As is typical with standards, not all standards have referred to these methods by the same name. Here is a table of what they have been called in the most used standards. The latest version of TIA-526-14 refers to them as one-cord, two-cord and three-cord reference methods. Some vendors use the term "jumper" instead of cable.

Reference Method	OFSTP-14 (TIA-526-14)	OFSTP-7 (TIA-526-7)	IEC 61280-4-1	IEC 61280-4-2
	Multimode	Singlemode	Multimode	Singlemode
One-cable	Method B	Method A.1	Method 2	Method A1
Two-cable	Method A	Method A.2	Method 1	Method A2
Three-cable	Method C	Method A.3	Method 3	Method A3

One-Cable Reference Test

The first option is using only 1 reference cable, the launch cable attached to the source. The meter measures the output of the launch cable at its connector and that power level can be set as "0 dB." Most power meters have a function that allows setting a 0 dB reference level at any input power by pushing a button on the front panel, so it is assumed that the meter is now set to read "0 dB."

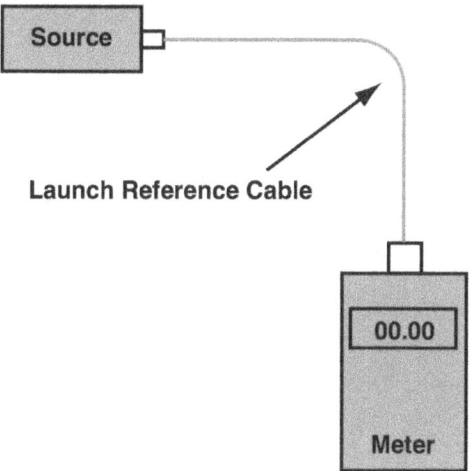

Setting "0 dB" reference with one cable

This method is preferred by most standards but only works with a test source and power meter that have connector interfaces that mate to the connectors on the cable plant being tested. This is less a concern for the source since a hybrid launch cable can be used with a connector on one end that mates to the source and a connector on the other (launch) end that can mate to the cable plant. Most meters have modular adapters that allow changing to an adapter that fits most connector types.

Note: *In all these test setups, it is assumed that the connection to the power meter is "lossless" because the detector in the power meter is large enough to capture all the light exiting the fiber, so it can be ignored when calculating the loss during reference setup and testing.*

Connector matching is more likely to be a problem with an OLTS that has fixed connectors that cannot be adapted except by hybrid cables, the reason that the other reference options are provided. That will be covered next.

If the output of the launch cable is set to 0 dB, what the meter will measure is the "dB loss" of the connection of the launch cable to the cable under test at point "1," the "dB loss" of the cable to test and the "dB loss" of the connection "2" to the receive cable attached to the power meter. Thus, this setup tests the loss of the entire cable being tested including the connectors on each end.

The meter now reads the dB loss of the test setup, which we call the loss of the cable under test.

The sum of these three "dB losses" measured by the power meter is what we call the loss of the cable plant under test. With the one cable reference, the value measured is what would reasonably be called the loss of the cable plant and the loss of the components in the cable plant including the loss of the connectors on each end.

The One-Cable reference method has another advantage over all other methods; it allows one to check the condition of the connectors on the reference cables directly to ensure they are of high quality. Simply connect the output of the launch reference cable (set to a 0 dB reference) to the receive cable to measure the loss of the connection.

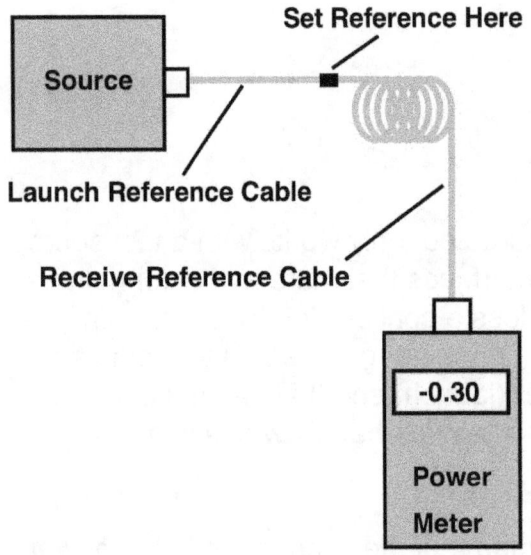

Testing reference cables to verify the condition of connectors

The loss of this mated pair of connectors should be around 0.3 dB or less generally and the cables replaced when the loss exceeds 0.5 dB. This step is important because the large number of matings these connectors endure when testing cables can cause wear and/or damage to the end faces of the connector ferrules that will result in high losses when testing other connectors. Generally the larger number of matings a connector gets, the higher the loss will be due to wear on the mating surfaces.

To reduce wear and damage, the connectors on reference cables must be properly cleaned and inspected often. Reference test cables should be inspected at least as often as every few tests, cleaned and then inspected again to determine if they are dirty or have been scratched or scuffed. When a connector shows some wear, test it as shown above to determine if it should be replaced.

Two-Cable Reference Test
If you are using an OLTS with fixed fiber optic connector interfaces, often SC or ST connectors, and need to test a cable plant with LC connectors, you may need to use hybrid reference test cables to mate to the connectors on the cable plant under test. This situation can also arise if you are testing connectors that do not have a matching adapter for the power meter. Since the output of the launch reference cable has a different connector than the meter input of the OLTS, you will need to set a 0 dB reference using hybrid cables on both the source output and meter input of the OLTS, creating a Two-Cable reference.

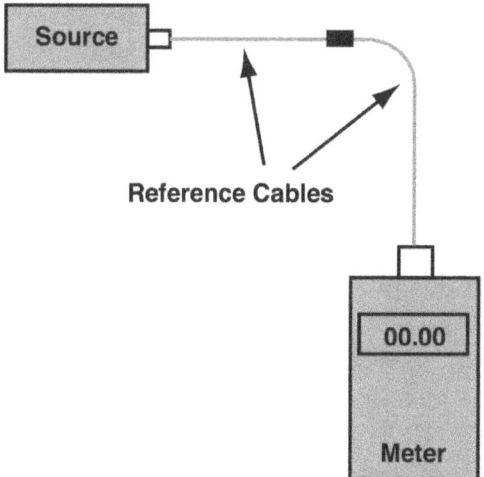

Setting "0 dB" reference with 2 cables

When you set the meter or OLTS to read "0 dB" with this reference setup, the reference includes a connection between the connectors on the launch and receive cables. The output of the launch cable is reduced by the loss of the connection to the receive cable, so when you disconnect the two reference cables and insert a cable to test, you are measuring the loss of the cable under test *less the loss of the connection included when setting the loss*.

What the power meter in the OLTS measures is:

   Loss of launch cable connection
\+ Loss of components in the cable plant
\+ Loss of receive cable connection
<u>- Loss of connection between reference cables when 0 dB reference was set</u>
= Total loss in dB, reduced by the reference cable connection loss

Therefore the measured loss of the cable plant will be lower than if it were measured with a One-Cable reference, reduced by the loss of the connection on the reference cables when setting the 0 dB reference. Furthermore, since that connection loss cannot be determined, it adds to the uncertainty of the measurement of loss.

This method of testing is generally the least preferred method of testing cables but has been included in instructions by many test equipment manufacturers, especially those selling OLTS for telecom cable plants. Over many years, telecom cable plants have had many different types of connectors (Biconic, D4, FC, Diamond, etc. as well as the more common SC), making it more likely that one will encounter a connector that is different from the one on the test set. And if the cable plant is telecom, there is a good likelihood of it having relatively high loss, 10, 20 dB or more. Thus subtracting an unknown connector loss of ~0.5 dB is of little consequence.

However, if one is testing multimode fiber in a data center or high speed LAN backbone where the loss is likely to be around 2 dB, an uncertainty of 0.5 dB is highly significant. Therefore this method is not a preferred method for applications where loss is lower and errors more significant.

Note: *This method also makes it difficult to test the quality of the reference test cables. If a microscope is available with an interface to the connectors on the cable plant being tested, they should be inspected carefully after cleaning. Since the OLTS or power meter cannot mate to the output of the launch cable, it is impossible to test the loss of the mated connectors. This unknown loss adds to the measurement uncertainty of this method. The relative measurement uncertainties of the different methods of setting a 0 dB reference will be covered later in this chapter.*

Three-Cable Reference Test
When testing connectors that are "plug and jack" or "male and female," the test setup gets more complicated. With these connectors, the permanently installed cable plant will have one type of connector on patch panels or wall outlets and the opposite gender on patchcords. Generally the patchcords

have the "plug" type of connector and the installed cable plant will have "jacks" on the patch panels or outlets. Since the connectors on the cable are plug and jack and can only be mated to the opposite gender and cannot be mated to the test equipment, a three-cable reference is used.

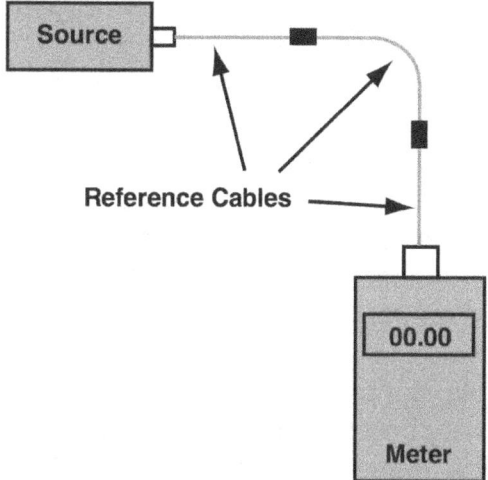

Setting "0 dB" reference with 3 cables

In this method, the launch and receive cables will be hybrid cables with one end terminated with a connector compatible to the test equipment and the other end terminated with a plug type connector compatible with the cable plant. Since the two plugs cannot be mated to each other, a third reference cable terminated with jack style connectors on each end is placed in between the two reference cables. This becomes the 0 dB reference for testing.

When testing cables, the third reference cable is removed and replaced with the cable to be tested. Another name for this test is a cable substitution test since the cable under test is substituted for the reference cable. The measured loss is the difference between the loss of the reference cable and the loss of the cable under test.

When the 0 dB reference is set with three reference connectors, there are two connections included. So when you remove the third reference cable and insert a cable to test, you are measuring the loss of the cable under test *less the loss of the two connections included when setting the 0 dB loss reference.*

What the power meter in the OLTS measures is:

  Loss of launch connection
 + Loss of components in the cable plant
 + Loss of receive connection
 - Loss of 2 connections between reference cables when 0 dB reference was set
 = Total loss in dB, reduced by the loss of the 2 reference cable connections

Therefore the measured loss of the cable plant will be lower than if it were measured with a one or two-cable reference because of the loss of the 2 connections on the reference cables included when setting the 0 dB reference. Furthermore, since those connection losses cannot be determined, it adds to the uncertainty of the measurement of loss.

The Three-Cable reference has another potential problem that can be confusing. If the 3rd reference cable is not very good or the connections to the other reference cables are either bad or dirty, the replacement with a cable to test can show quite large errors, even sometimes showing a gain. When using this method, it is very important to use only high quality cables and be careful to clean and inspect all connectors on reference cables to minimize measurement errors.

Note: Some instructions and even standards claim that the Three-Cable reference method tests the cable under test but excludes the loss of the connections on each end because those connections are included in the setting of the 0 dB reference. While this appears to be a logical assumption since two connections are indeed included in the setting of the 0 dB reference, the assumption is not correct due to the uniqueness of fiber optic connections. Fiber optic connections are defined as the optical connection between two connectors. For most types of fiber optic connectors that connection includes a separate mating adapter that provides the alignment of the two connector ferrules. The loss is a function of the physical characteristics of both of the connectors and the mating adapter that aligns them. The precise alignment required for a fiber optic connection is on the order of microns or less – a micron is only one-millionth of a meter – and the actual loss of a mating can vary if two connectors are mated repeatedly. When the third reference cable is removed from between the launch and receive reference cables and replaced with the cable to be tested, the loss is highly likely to be different due to differences in the fiber and connectors of the cable being tested. Therefore one is not including the connections to the test cable in the 0 dB reference but two different connections whose loss will be deducted from the final test results. Assuming this method tests the cable without ending connections is technically incorrect and as the section below on "Measured Loss With Each Reference Method" shows, the uncertainty of the Three-Cable reference method is much higher than the single cable method that does include the connection losses from both ends of the cable.

Example With MPO Connector
The MPO connector is a type of plug and jack connector that aligns connectors with metal pins on one connector and holes for those pins on the other connector. Connectors are mated in adapters that snap onto the connector to hold it in place while the pins align the ferrules. The "jack" on a patch panel or outlet will have a connector with pins in an adapter because the pins are protected from damage inside the jack. Patchcords will have connectors with holes because the protruding pins can easily get damaged in handling. In addition, there is a key on MPO connectors that can be aligned either up or down to determine fiber polarity. Fortunately, manufacturers now make MPOs with retractable pins and switchable keys that simplify testing by allowing the connector on a reference test cable to be adapted to any style of MPO connector.

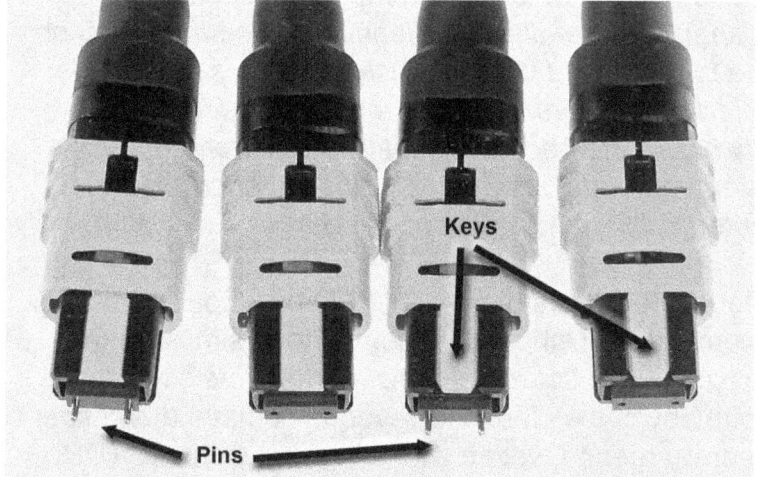

MPO connectors (L > R), Key down pins, holes, Key up pins, holes

The MPO connector has 12 fibers aligned in a row with options for 1 to 6 rows of fibers. Most test equipment is designed to test a single fiber at one time. Testing MPO connectors is further complicated by the need for a breakout cable that has MPO connectors on one end and multiple single fiber connectors compatible to the test equipment on the other.

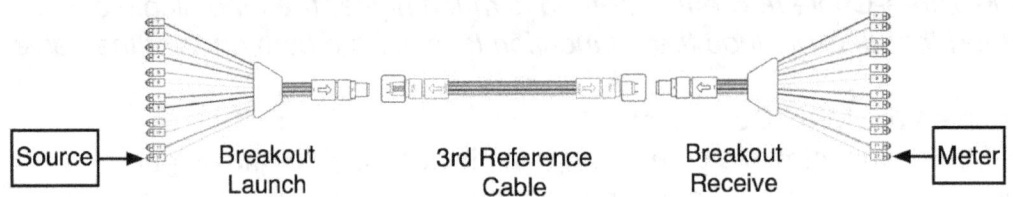

Reference test setup for MPO-MPO cable test using 3-cable method

The reference setup for the MPO connector requires two breakouts and a short reference cable that simulates the cable that will be tested. The source and meter are attached to one fiber of the cable at each end and the loss tested for that fiber.

Note: As with the two-cable reference method, this method also makes it difficult to test the quality of the reference test cables. If a microscope is available with an interface to the connectors on the cable plant being tested, they should be inspected carefully after cleaning. Since the OLTS or power meter cannot mate to the output of the launch cable, it is impossible to test the loss of the mated connectors. This unknown loss adds to the measurement uncertainty of this method. The relative measurement uncertainties of the different methods of setting a 0 dB reference will be covered later in this chapter.

If you need to test a prefabricated cable plant with MPO backbone cables and breakout cables to conventional single fiber connectors, similar to the drawing above, using a regular light source and power meter with reference cables can be used as with any cable plant. However if you have a bad fiber in the link, locating it generally means testing with a high-resolution OTDR. Even those cannot resolve the short fiber in a breakout module so at most you will be able to isolate the problem to the module itself, which will need replacement.

Measured Loss With Each Reference Method
To illustrate the differences in loss measured with each reference measurement, tests were performed with a simulated cable plant using the same source and power meter with 10 sets of high quality launch and receive reference cables to determine the variation in loss measured. Each set of reference cables were tested against each other to ensure they had low loss and the simulated cable was tested in both directions. Here are the results:

Reference Test Method	Results, loss and standard deviation in dB
One-Cable reference	2.96 dB, ±0.02 dB
Two-Cable reference	2.66 dB, ±0.20 dB
Three-Cable reference	2.48 dB, ±0.24 dB

Note how the loss of the cable tested reflects the comments we made above. The one cable reference method has higher loss than the other methods, but it also has much less measurement uncertainty (standard deviation.) The two- and three-cable reference methods have less loss because we have subtracted the connector loss(es) included when we set the reference for 0 dB loss. The uncertainty is higher because of the uncertainty created by the unknown losses of the connections of the reference cables when setting the 0 dB reference.

Which Reference Method Is The "Correct" Method?
Technically, all of them are correct as long as the test method is included in the documentation of the test. Standards generally allow different methods of setting the 0 dB loss but require reporting the method used to be recorded along with the test results. Having the test method as well as the test data allows the proper evaluation of the measurement.

Testing Reference Cables

The "single ended" test is generally used with short cables like patchcords. The test only includes the connector mated to the connector on the launch reference cable. The other connector is connected directly to the power meter that collects all the light from the end of the fiber. The power meter end effectively has no loss. This test is mostly used to test short cables like patchcords where the loss of the fiber is insignificant and there are no intermediate connections. Since it allows the measurement of the connectors on each end separately, it is a more thorough method of testing these short cables.

The single ended test is performed without a receive reference cable. The "0 dB" reference measurement is made with the test source with launch reference cable and the power meter using the same One-Cable method. The cable under test is mated to the connector on the launch cable and the power meter. This test requires a power meter with an adapter compatible with the connectors on the cable to test. It can also be done with an OLTS if the connectors are compatible.

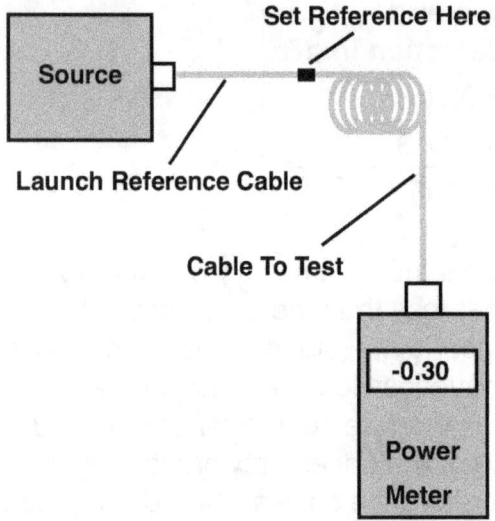

Insertion loss test diagram – single-ended test

The output of the launch cable is the 0 dB loss reference, so for short cables like patchcords, the measurement is basically the loss of the connection between the launch cable and the test cable. The cable to test can be reversed and the connector on the other end tested separately.

This test is used by manufacturers making cables on patchcords because it allows testing each connector individually, ensuring that both connectors meet specifications. Data provided on patchcords normally includes the loss tested in each direction. If the cable were tested with the double ended test, one would measure a total loss of both connectors and could not identify if one connector were out of specification.

Test Conditions For Accurate Insertion Loss Testing

As with any testing, to reduce measurement uncertainty, it is important to consider and as, far as possible, control test conditions. It is important to understand all the contributions to measurement uncertainty and how their effects can be minimized by the test operator. Many sources of measurement errors are not controllable by the user. They depend on the manufacturer of the test equipment, optical fiber, connectors, etc. and their quality control.

We will summarize the most important issues for reducing measurement uncertainty here and go into it in more detail in Chapter 10.

The first and most important issue is to decide which reference method to use (1, 2 or 3 cables) and record that with the test data. Since the test results change considerably with the choice of reference method, this is very important information to record. After setting the "0 dB" reference, it is important to not disconnect the reference launch cable from the source. The connection between the source and cable can change if it is disconnected and reconnected invalidating the "0 dB" reference set.

Secondly, monitor the type and condition of all the test reference cables. This is probably the biggest cause of random errors in testing. Of course the fiber in the reference cables must match the type in the cables being tested and the connectors must be compatible. The condition of the connectors on the reference cables should be inspected for cleanliness and damage, cleaned and inspected again to ensure proper cleaning. Then the reference cable connectors should be tested against each other to ensure they are still low loss. As the loss increases from more testing, replace the cables or refurbish the connectors by repolishing using diamond film.

The connections to test cables have another important component, the mating adapter used with connectors. All the single fiber connectors use mating adapters to align the ferrules. These adapters are available in several types, depending on the alignment sleeve for the ferrules. Some inexpensive mating adapters use plastic alignment sleeves that should never be used for testing. These sleeves wear out quickly and leave dust and residue on the connectors. The acceptable mating adapters use metal or ceramic alignment sleeves. Metal sleeves work for hundreds of tests but ceramic sleeves will last many times longer and leave no residue.

Test equipment used for insertion loss testing should be checked and power meters calibrated regularly, according to manufacturers specifications. When making tests, the biggest problems are with sources. First of all, the source needs to be stable; if its output power drifts, the 0 dB reference will be lost and all tests will be wrong. You can test your own source by connecting it to a meter with a patchcord and turning it on. Note how long it takes the output to be stable and use that time as a warmup time when setting up to make tests. If the source is not stable over time, varying more than 0.1 dB after warmup, it should be checked or replaced with a more stable source.

The second issue with sources is modal conditioning. This is mainly a multimode fiber and LED test source problem, but even single mode sources with lasers can have modal problems. The user should ensure that multimode launch cables have proper mode power distribution since that can affect the loss measured significantly. Using a simple mandrel wrap and checking the modal distribution with a HOML (higher order mode loss) test as described below will greatly reduce measurement errors. For singlemode fiber, a simple loop mode filter is all that is needed. With all reference cables, be careful to not stress them during the tests as that can induce loss that will change the 0 dB reference and or create changes in the modal distribution.

The technician performing the tests should be experienced in the process, familiar with the procedures and conduct every test in the same manner. Even small issues like stress on reference cables can make big differences in measurements.

If reasonable precautions are taken, what is the likely accuracy of loss measurements? Experience has shown that typical measurements have an uncertainty of approximately ±10% of the measured value in dB. Thus a 2dB loss has an estimated uncertainty of ±0.2 dB.

Modal Distribution In Multimode Fiber
In multimode fibers, different types of sources launch with different modal distribution. A LED typically has a wider light output that overfills a fiber, while lasers, even VCSELs, underfill the fiber. This also produces conditions in the fiber that affect fiber attenuation measurements and connection measurements.

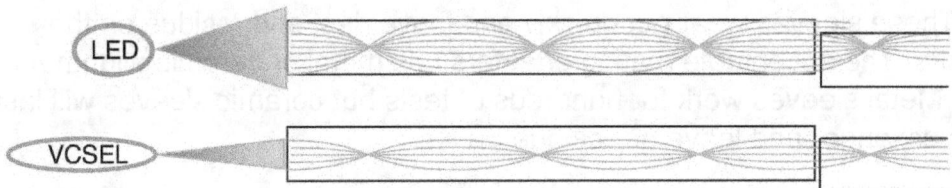

Modal distribution from LED (overfill) and VCSEL (underfill) sources affects fiber attenuation and connector loss

Controlling launch conditions are important to making accurate measurements of multimode fiber. In order to control modal conditions, it is first necessary to be able to measure it.

The most effective and widely used method of mode conditioning is the mandrel wrap, tightly coiling the launch cable around a specific sized mandrel or rod. The tight bends causes stress on the fiber that produces loss, primarily in the higher order modes. The stress also scrambles lower order modes, making the mode fill more consistent.

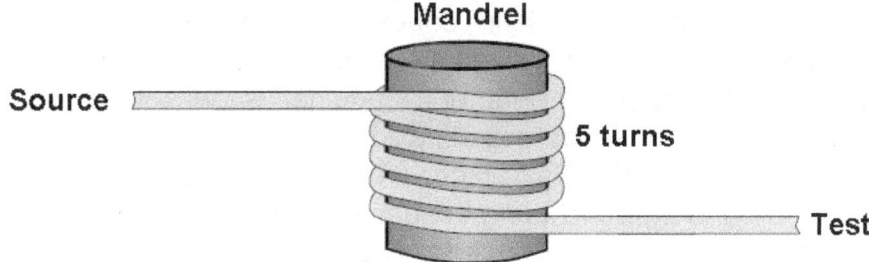

Mandrel wrap mode controller

The mandrel wrap mode controller is based on five turns around a specified size mandrel, but the mandrel size varies according to the cable or fiber type. Below are the TIA standards for the mandrel wrap.

TIA Specified Mandrel Size – Wrap 5 Turns				
Fiber/Cable Type	3mm Jacket	2 or 2.4 mm Jacket	1.6 mm Jacket	900 micron buffered fiber
50/125 micron	22 mm	23 mm	24 mm	25 mm
62.5/125 micron	17 mm	18 mm	19 mm	20 mm

Mandrels are available from test equipment manufacturers or can be made from readily available materials such as a wooden dowel or plastic rod. The 22mm mandrel is very close to 7/8 inch.

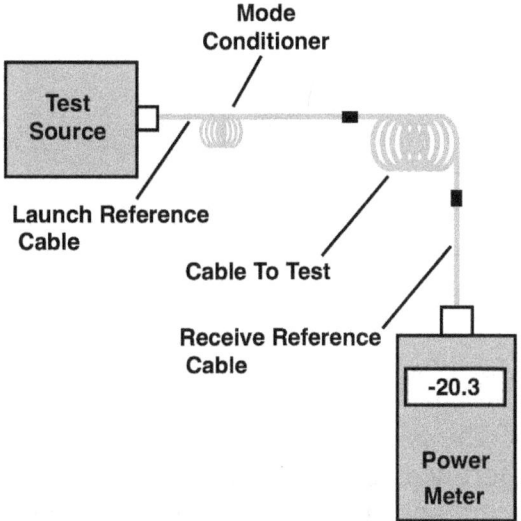

The mandrel wrap mode conditioner should be made on the launch reference cable near the test source

As most multimode systems have migrated to 850 nm VCSEL sources, a new standard for modal distribution was introduced called encircled flux (EF.) Encircled flux was a theoretical model of modal distribution intended to simulate an ideal VCSEL light output pattern. It's original use was in modal bandwidth simulations as VCSEL sources allowed multimode fiber systems to operate at 1 Gb/s and later 10 Gb/s or higher. These systems had very low power budgets, around 2 dB including the penalties for dispersion, so it became more important to reduce the uncertainty of loss measurements.

Encircled flux has been adopted by most insertion loss testing standards for laser optimized 50/125 micron fibers today. TIA in the US specifies EF testing for OM3 and OM4 fiber at 850 nm only, considering the fact that legacy systems using older fibers like OM1 or OM2 were tested under prior standards. Some ISO/IEC standards may call for EF testing for all fibers.

Creating And Measuring EF Launch Conditions
In order for a standard like EF to be usable, it must be easily created and tested. Much work was done to understand EF and how to create it. A few expensive commercial devices have been available but the conclusion of the TIA standards committee is that the mandrel wrap will create EF modal conditions. In the TIA, a test was developed and proven to be acceptable in determining if a launch met EF conditions. The test is called "higher order mode loss" or HOML.

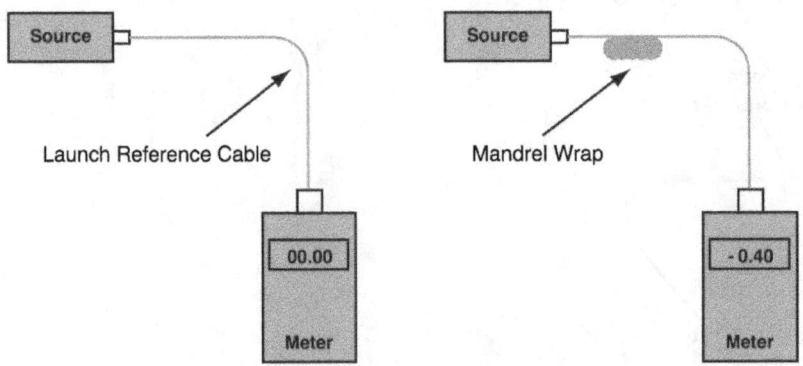

Higher order mode loss (HOML) test for EF launch conditions.

HOML testing is extremely easy. Connect the launch reference cable to a source and measure the output of the reference cable with a power meter. Wrap the launch reference cable around the specified mandrel and measure the output again. There are three options when analyzing the test results.
  • If the measured power is reduced by 0.20 to 0.60 dB, the source is essentially EF compliant and ready to use, without the mandrel.

Remove the mandrel and make your tests.
- If the HOML is >0.60 dB, leave the mandrel on the reference launch cable and make measurements.
- If the HOML is <0.20 dB, the source has too low a mode fill and should not be used.

Conclusion On Multimode Modal Control
The recommended procedure is to always use the mandrel wrap on the launch cable and use a HOML test to verify the launch conditions. Include these test conditions in the documentation of all testing.

Singlemode Fiber Mode Control
Singlemode fiber only supports a single mode, but when laser sources are coupled into a singlemode fiber, there may be several modes coupled for a short distance. When singlemode fiber is tested with a laser source a mode filter should be included in the launch cable to remove any higher order modes. All that is needed is a small loop in the launch cable near the source with a diameter of 30-60 mm (1.2 to 2.4 inches.) The loop should be taped to make it stable for consistency.

Note: Other factors in measurement uncertainty for insertion loss measurements will be covered in Chapter 14 on metrology and measurement errors.

What Loss Should You Get When Testing The Loss Of Cables?

Testing provides data but data require interpretation. After all the care that goes into setting up a test and executing it properly, it is necessary to decide if the test shows the component loss, e.g. the loss of a cable plant, is within expected limits. With loss testing, the data is generally compared to a loss budget, a calculated estimate of the loss of the cable plant.

The network designer should calculate a loss budget for all the cable plants to be tested before testing, preferably during the beginning of the design phase. The loss budget is initially used to ensure the planned communications system will work on the cable plant as designed. In addition, the loss budget will provide estimates for the expected measurement results. This will provide reference loss values to compare test results against and help make pass/fail decisions on each fiber in the cable plant.

A loss budget estimates the loss of a fiber optic cable by adding together the estimated loss of each component in the cable – fiber, connections, splices and any other devices that cause loss like switches or splitters used in passive optical networks.

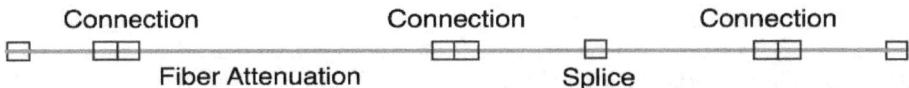

Components included in loss budget

In the drawing above, the fiber loss over the entire length of the cable, the one splice, and 5 connections are included. There are 5 connections because there are obviously 3 connections in the cable shown *plus* connectors on each end. In order to test this cable plant, it will be connected to reference cables on each end, so each of the connectors on the end of the cable will become connections that must be included in the loss budget.

The loss budget is calculated as:

	Fiber loss = fiber length (km) X attenuation coefficient (dB/km at test wavelength)
+	Connection loss = Number of connections (see Note) X loss/connector (dB)
+	Splice loss = Number of splices (including ends) X loss/splice (dB)
+	Any other loss devices
=	Loss budget

Note: Loss budgets are generally calculated including the connectors on each end of the cable. That is the proper way to calculate loss budgets when testing with the one-cable reference method that is the most widely used method. If you are using a two-cable reference method, you will include one connection in the reference so only one end connection should be included when calculating the loss budget to compensate for that reference method. If you are using the three-cable reference method, neither end connector should be included in the calculation because two connections are included in the three-cable reference.

A loss budget is only as good as the assumptions used for component losses. Some designers think that one should use component losses from industry standards, since they are standards. But the values for component losses in standards are worst-case values, negotiated in the standards making process to be high enough that few components will fail testing. For example, the TIA standard for connector loss is 0.75 dB, a value that was chosen because it would accommodate any connector including the high loss multifiber MPO connector.

While it is difficult to generalize about component losses for loss budgets, here are some realistic guidelines:

- Connections:
  - 0.3 to 0.5 dB loss for adhesive/polish connectors,
  - 0.5 to 0.75 for prepolished/splice connectors
  - 0.5 to 1.0 dB for MPO connectors
- Splices:
  - 0.2 dB to 0.3 dB for fusion splices on multimode fiber
  - 0.05 dB to 0.2 dB for fusion splices on singlemode fiber
  - 0.2 dB to 0.3 dB for mechanical splices
- Optical fiber attenuation
  - 3 dB/km for multimode fiber at 850 nm
  - 1 dB/km for multimode fiber at 1300 nm
  - 0.5 dB/km for singlemode fiber at 1310 or 1550 nm (premises tight-buffer cable)
  - 0.4 dB/km for singlemode fiber at 1310 nm (OSP cable)
  - 0.25 to 0.3 dB/km for singlemode fiber at 1550 nm (OSP cable)

It is important to understand that a loss budget is a calculated estimate of the loss of a cable. The "estimate" part comes from the values chosen for the attenuation coefficient of the fiber and the loss of connectors and splices. For premises cabling, you can choose the TIA 568 values that are "worst case" and quite high - connectors are 0.75 dB for example - or you can choose something more realistic.

To illustrate the difference, a 850nm link 200 meters (0.2 km) long with OM3 multimode fiber having 2 intermediate patch connections (4 connections with the ends) would have a loss of 3.7 dB with TIA values and 1.8 dB with typical values – that's more than a factor of 2!

	Typical values	TIA (worst case)
Fiber	0.2 km X 3 dB/km = 0.6 dB	0.2 km X 3.5 dB/km = 0.7 dB
Connections	4 X 0.3 dB = 1.2 dB	4 X 0.75 dB = 3 dB
Total	1.8 dB	3.7 dB

If you are comparing loss budget estimates to test data it is obviously important to use realistic estimates, otherwise the likelihood of passing cables that should fail is very high.

Note: FOA offers a free loss budget calculator for iPhones on the Apple iTunes Store. There are also available Android apps that calculate loss budgets.

Interpreting Optical Loss Measurement Data

Earlier in the chapter we discussed making insertion loss measurements with a light source and power meter and the causes of errors in those measurements. The estimate was that these measurements generally had potential errors of as much as +/-10%, a combination of all systematic and random errors.

An error of up to 10% is not insignificant. That's important when interpreting data from a power meter display. For example, a cable that measures 3.17 dB loss might actually have a loss of ~2.85 to 3.50 dB. That meter reading has a resolution much higher than needed for the measurement - 0.1dB is plenty for most loss measurements - and can falsely indicate that the measurement is more accurate than it really is.

The confusion begins when deciding what to do with that measurement. Do you pass or fail that cable? Does it meet the limits calculated in a loss budget? Will it work with a system that has a power budget of 3dB?

A power budget is the maximum loss specified for a specific fiber optic network, say an Ethernet LAN link. See Chapter 9. Theoretically, the power budget can be calculated as the difference between the output power of the transmitter and the required input power at the receiver.

However there may be other issues to consider. When it comes to multimode fiber networks, the bandwidth of the fiber affects the power budget for high speed systems. Simply looking at the transmitter and receiver specs for a 10G LAN, you might surmise that a 5 dB power budget would be required. But when you factor in the bandwidth of the fiber that creates a power penalty, you will find that a 200m link must have less than 2 dB loss.

So let's look at the cable we said measured 3.17 dB loss. It's 200 meters long and has 4 connections including the connections on each end – that's the cable we calculated a loss budget for in the previous section. What happens when we compare our loss budget? Does this cable pass or fail? Here is the data we are faced with:

Measured loss: 3.17 dB, probable actual loss 2.85 to 3.50 dB

Loss budget: TIA - 3.7 dB, typical - 1.8 dB

Power budget: Low speed - 5 dB, high speed 2 dB

The 3.17 dB loss tested is within TIA limits but way over the typical estimate. It's OK for low speed networks but way too much loss for high speed networks.

That cable with a loss of 3.17 dB should be a "FAIL." The biggest concern is connector loss. Using the TIA estimate, subtract the 0.6 to 0.7 dB loss of 200 meters of fiber and you get about 2.5 dB loss from connectors. If that cable has four connectors, that's more than 0.6 dB each on the average, or if 3 connectors are more typical (~0.4 dB,) one connector has about 1.3 dB loss. That means its time to start troubleshooting.

Another common decision is what do you do if the loss budget is 1.8 dB and the cable measures 1.95 dB loss if it's a cable for a 10G system with 2 dB margin? Is that a failure? No it's a "PASS." The measured value and the loss budget are both estimates and the two values are within +/-10% - reasonable margins of error.

Remember the measured data has some error and the loss budget is an estimate. It is necessary to consider all the possible errors and use your judgment. Now you need to troubleshoot that 3.17 dB cable.

Troubleshooting Hints

Most problems with high cable loss are caused by bad or dirty connectors, high loss splices or stress loss in the cable caused during installation.

The first step is connectors should be inspected with a microscope for dirt, scratches cracks, or other damage and thoroughly cleaned. Visual fault locators can check for continuity, proper connections and, if the cable jacket permits, high loss bends or breaks.

If you have high loss in a single cable with connectors on each end, you can reverse it and test in the opposite direction using the single-ended test method. Since the single ended test only tests the connector on the end connected to the launch cable, you can isolate a bad connector this way. The bad connector is the one at the end mated to the launch cable when you measure high loss.

High loss in the double-ended test should be isolated by retesting single-ended and reversing the direction of test to see if the end connector is bad. If the loss is the same, you need to either test each segment separately to isolate the bad segment or, if it is long enough, use an OTDR.

Documentation

Like every step of the fiber optic design, installation and operation processes, it is important to fully document the test and record all relevant data. All tests should be recorded with the following data as a minimum:

For all tests performed at one time:
- Date of the test
- Location of the test
- Environmental conditions (temp/humidity/local conditions)
- Cable plant identification (cable type/fiber type/connector type/ length)
- Type of test (visual inspection, insertion loss, OTDR, CD, PMD, SA)
- Test equipment used (type, brand, model, serial number, date of last calibration)
- Wavelength
- Insertion loss: reference method (1/2/3 cable reference methods)
- OTDR: manual or auto test
- Criteria for pass/fail (loss budget criteria)

For each individual test
- Identification of component under test (e.g. fiber #)
- Test results (actual data and pass/fail if noted)
- Note if results are filed electronically

Chapter Exercises

- Set up a power meter and reference cable to measure the output power of a test source (set 0 dB on the meter) and measure the warmup time for as many test sources as you can. See how long they take to become stable to better than 0.1dB. Then leave them on for 30 minutes and see if they remain stable.
- With the source, reference cable, power meter setup, note the power (set 0 dB on the meter) then disconnect the source and reconnect several times to see how much it varies in power level.
- Measure the loss of a long cable plant using one-, two- and three-cable reference methods and not the differences.
- Make a MM insertion loss test with and without a mandrel wrap and

compare results.
- Test a reference cable with the HOML test to see if your source meets standards.
- Test fibers with mismatched reference cables (50/62.5 and 50/SM) and compare results

Chapter Quiz

1. Cables tested with an OTDR do not require insertion loss testing with a source and meter or OLTS.
   True
   False

2. OTDR testing will generally give the same test results as insertion loss testing with a light source and power meter.
      True
      False

3. 5. What test instrument(s) are used for insertion loss testing?
   A. OLTS or power meter and test source
   B. VFL
   C. OTDR

4. What is a "0 dB" reference?
   A. The output of the test source
   B. The connection loss of the source and launch cable
   C. The power level measured during the one-, two- or three-cable reference setting process
   D. What the power meter measures

5. Multimode graded-index glass fiber optic cables are tested with sources at _____ and _____ wavelengths.
   A. 650, 850 nm
   B. 850, 1300 nm
   C. 980, 1400 nm
   D. 1310, 1550 nm

6. Singlemode graded-index glass fiber optic cables are tested with sources at _____ and _____ wavelengths.
   A. 650, 850 nm
   B. 850, 1300 nm
   C. 980, 1400 nm
   D. 1310, 1550 nm

7. What type of source is used for testing singlemode fibers?
   A. LED
   B. VCSEL
   C. Laser

8. What type of source is used for testing multimode fibers?
   A. LED
   B. VCSEL
   C. Laser

9. What type of source is not recommended for testing multimode fibers even though it is used as a network transmitter?
   A. LED
   B. VCSEL
   C. Laser

10. How many methods are included in standards for setting the "0 dB" reference for insertion loss testing?
   A. One
   B. Two
   C. Three
   D. Four

11. Reference cables must match the _____ of the cables being tested.
   A. Fiber size and type
   B. Fiber size and connector type
   C. Connector type
   D. Fiber size and loss specification

12. The reference cables needed for testing insertion loss _____.
   A. Can be any old cables in your toolkit
   B. Should be random patch cables used for connecting equipment to the cable plant
   C. Should be known good cables regularly tested for low loss
   D. Must be special reference-grade test cables purchased only from the test equipment manufacturer.

13. If a cable plant tests shows unacceptably high loss, the first thing the tech should do is _____.
   A. Inspect and clean all connectors
   B. Check the calibration of the test equipment
   C. Test the reference cables for connector loss
   D. Test again with a high resolution OTDR

14. Loss budgets are calculated by adding up _____ from the cable plant.
   A. All connector losses, including the ones on the end of the cable
   B. All splices
   C. All fiber attenuation
   D. All of the above plus any other passive devices in the cable plant

Chapter 9

Optical Time Domain Reflectometer (OTDR) Testing

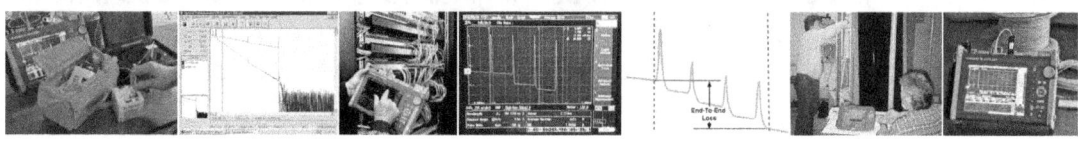

Objectives: From this chapter you should learn:
What an optical time domain reflectometer (OTDR) is
How OTDRs are used to test fiber optics
How the OTDR measures length, loss and reflectance
How to set OTDR test parameters correctly
What contributes to OTDR measurement uncertainty

What is OTDR Testing?

OTDR testing is a complement to insertion loss testing. While insertion loss testing simulates the loss of a cable plant with a light source and power meter, the OTDR uses an indirect test method that provides more but different information, including a "snapshot" of the entire cable under test. That snapshot, called a trace or signature, contains information on fiber length and attenuation, splice or connection loss, reflectance at splices or connections, losses caused by stress on the cable and total loss of the cable plant.

Portable OTDR

The Optical Time Domain Reflectometer (OTDR) is useful for testing the integrity of fiber optic cables and providing data for future reference. Traces taken at installation can be stored and if problems arise at a later time, comparisons can be made between the original trace and a new trace to find changes in the cable plant. Analyzing the OTDR trace is always made easier by having documentation from the original trace that was created when the cable was installed.

Most OTDRs are designed for testing long cables but some new instruments also allow testing short cables used in fiber to the home or data centers. OTDR testing should generally be used in addition to, not instead of, measuring insertion loss in the fiber optic cable. Some standards for cable plant testing now reference both insertion loss and OTDR testing but generally require insertion loss testing and allow OTDR testing as an option. The user of this instrument must know when its use is appropriate and how to optimize the testing for the application.

When And Where Is An OTDR Used?

The OTDR has primarily been used to test long OSP cables that are spliced together from numerous shorter cables. When these cables are fusion spliced, the splicing machine will give an estimate of the splice loss, but the OTDR allows actually seeing the results of the splice and confirming that it was made properly.

Besides verifying splice loss, the OTDR can measure the attenuation coefficient of the fibers in the cable and even find and locate places where a cable has been kinked, bent or stressed enough to cause loss or damage. If a cable is broken, the OTDR can locate the break from the end of the cable. If data from the installation is available, comparing that original data to current data helps in troubleshooting cable plant and communications network problems.

Using OTDRs in premises cabling is sometimes problematic. Most OTDRs are not good tools for short link testing. The way the OTDR works limits its distance resolution so that few instruments have sufficient resolution to distinguish the typical premises cabling patchcord from the cable plant it is plugged into. Some OTDRs which have been optimized for short links have the flexibility in setup to allow optimizing test conditions for short premises cables, but the operator needs to be knowledgeable in the operation of the OTDR to get valid data.

In recent years, OTDRs have been included in standards for testing cables and installed cable plants as an additional test method, not as a replacement for traditional insertion loss testing. These standards have included some new test methods and instructions on interpreting the data which have been included in this chapter.

How Does an OTDR Work?

Unlike sources and power meters that measure the loss of the fiber optic cable directly, the OTDR works indirectly. The source and meter duplicate the transmitter and receiver of the fiber optic transmission link, so the measurement correlates well with actual system loss. The OTDR, however, uses an optical characteristic of fiber to indirectly measure loss.

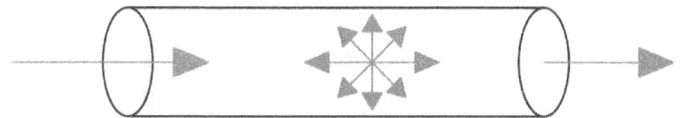

Scattering in an optical fiber

The biggest factor in optical fiber loss is scattering. In fiber, light is scattered in all directions, including some scattered back toward the source as shown here. The OTDR uses this "backscattered light" to make measurements. It also uses reflected light from fiber joints or cleaved fiber ends for measurements.

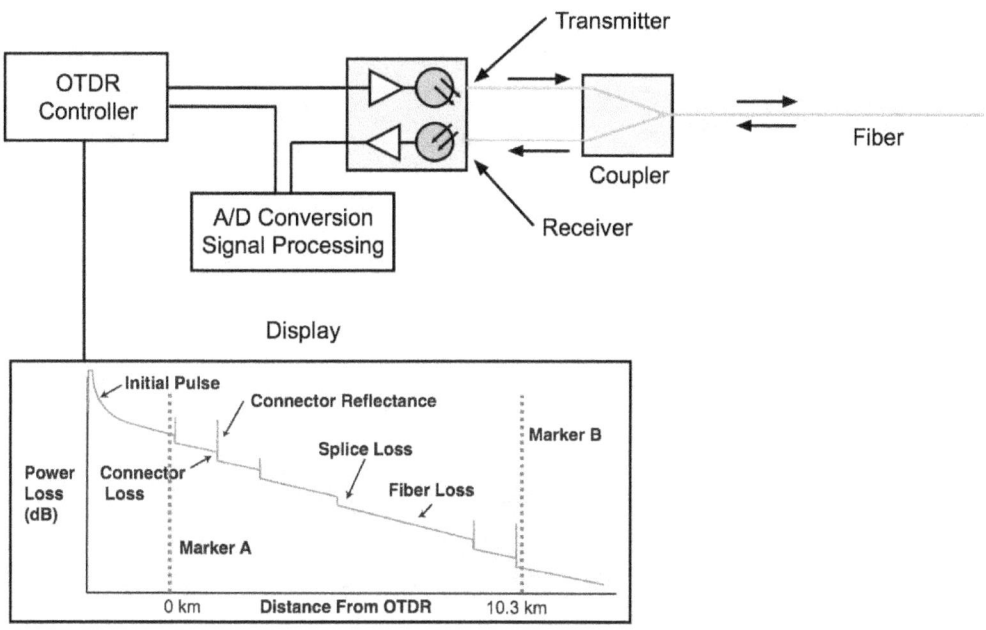

OTDR block diagram and display

The OTDR consists of a laser transmitter that sends a powerful pulse of light down the fiber. Backscattered light from the fiber and reflected light from connectors or other reflective components returns to the OTDR through the fiber. The returning light is directed to a sensitive receiver through a coupler in the OTDR front end that measures the light.

For each measurement, the OTDR sends out a pulse and measures the light coming back over time. At any point in time, the light the OTDR sees is the light scattered from the pulse passing through a region of the fiber. Think of the OTDR pulse going down the fiber as being a "virtual source" created by the backscatter that is testing all the fiber between itself and the OTDR as it moves back down the fiber.

To connect the OTDR to the cable under test, you generally use a launch cable to connect the OTDR to the cable under test and often a receive cable on the far end. The launch cable is fairly long to allow the trace to settle down after the overloaded caused by the test pulse from the OTDR (creating what is called a "dead zone".) The launch cable also provides a connection to the cable under test that allows testing the connector on the cable. To test the connector on the far end of the cable under test, you can attach a receive cable to it. The receive cable is also useful in other measurement functions such as bi-directional testing.

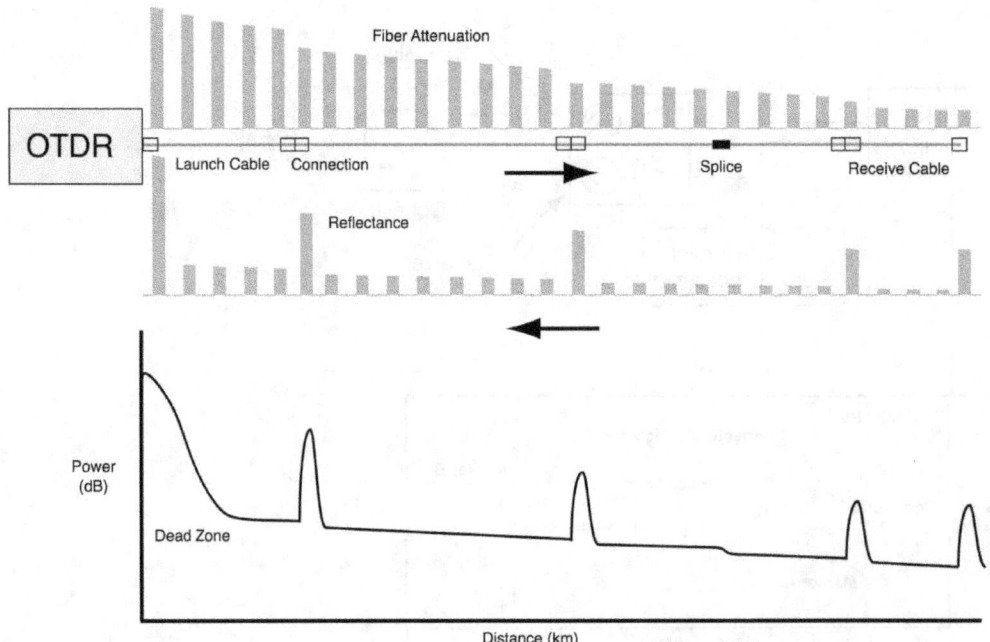

OTDR test and resulting trace

Since it is possible to calibrate the speed of the OTDR test pulse as it travels down the fiber from the index of refraction of the glass in the core of the fiber, the OTDR can correlate what it sees in backscattered light over time with an actual location in the fiber. Thus it can create a display of the amount of backscattered light at any point in the fiber along its length.

There are some calculations involved. Remember the light has to go down the fiber and come back, so you have to factor that into the time-distance calculations, cutting the time measured in half to get the distance down the fiber. One must also cut the loss in half, since the light sees loss in both directions. The power loss is a logarithmic function, so the power is measured and displayed in dB.

The amount of light scattered back to the OTDR is proportional to the backscatter coefficient of the fiber, peak power of the OTDR test pulse and the length of the pulse sent out. If you need more backscattered light to get good measurements, you can increase the pulse peak power, increase the pulse width or send out more pulses and average the returned signals. All three methods are used in most OTDRs, with user allowed control of some of the selections.

OTDRs are almost always used with a launch cable and generally use a receive cable on the far end. The high power of the test pulse coupled into the fiber connected to the OTDR overloads the receiver of the OTDR for a brief time, causing what is called a "dead zone." The launch cable, sometimes also called a "pulse suppressor," allows the OTDR to settle down after the test pulse is transmitted into the fiber and provides a reference connector for testing the loss of the first connector on the cable. Highly reflective events like connectors create smaller dead zones after each event also, and it is important to avoid these dead zones when making measurements. A receive cable should be used on the far end to allow measurements of the connector on the end of the cable under test also.

OTDRs Depend On Software
Most OTDRs today use a common PC operating system on which proprietary software controls the instrument and processes the measurements. Each OTDR's software is somewhat different. In talking to manufacturers, reading manuals and discussing OTDR operating methods with very experienced users, it seems that each instrument has slightly different systems.

This shows up in functions like the autotest mode where the parameters chosen by the instrument may be fixed or variable depending on operating conditions and the allowed time for making a measurement. Complex functions like LSA (least squares approximation) and reflectance measurements appear to use different methods and algorithms.

In evaluating OTDR performance, it is impossible to separate the effects of hardware and software operation. In our own tests, we have seen different OTDRs give different measurements and we cannot but wonder where the variations occur. In addition, OTDRs are very complex instruments. It can take days of work to understand how to use some functions, just like most other PC software. We highly recommend using the manufacturer's manuals to learn how to use your instrument properly and also use this book as a way of understanding how the OTDR can make measurements to ensure getting the most out of such a complex instrument.

Information in the OTDR Trace

The OTDR display ("trace" or "signature" as it is called) is full of information and data. Many OTDRs now use touch screen technology to make operating the instrument easier, so there may be touch buttons on the screen to choose OTDR operations. Here is a typical modern OTDR display.

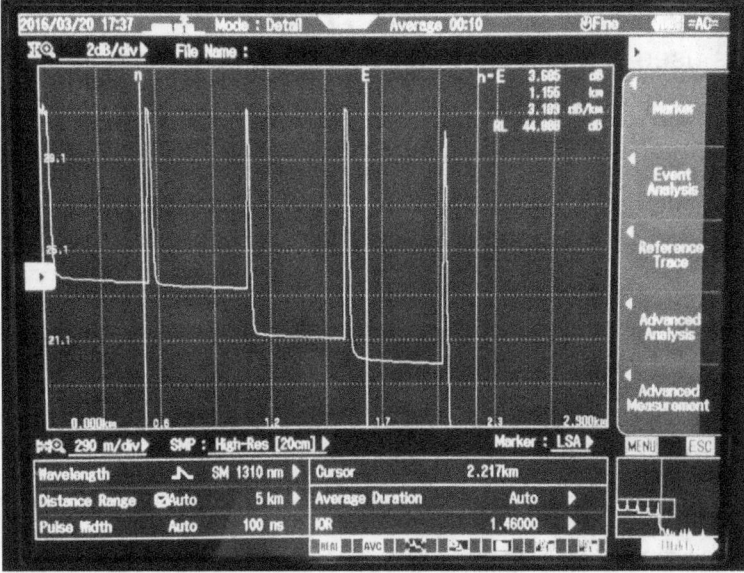

Typical modern OTDR trace with setup data below the trace and touch screen buttons on the right

The OTDR display shows the fiber trace in the center with data below that gives setup data or analyses of the trace and all events. The setup data shown above shows the test wavelength, OTDR range (distance for analysis,) pulse width, averaging duration, index of refraction (which relates to the speed of light in the fiber) and the location of the cursor.

After the OTDR analyzes the trace, which can generally be done either manually by the operator or automatically using OTDR software, the display will show an event table below the display that identifies events, gives their location and measurements.

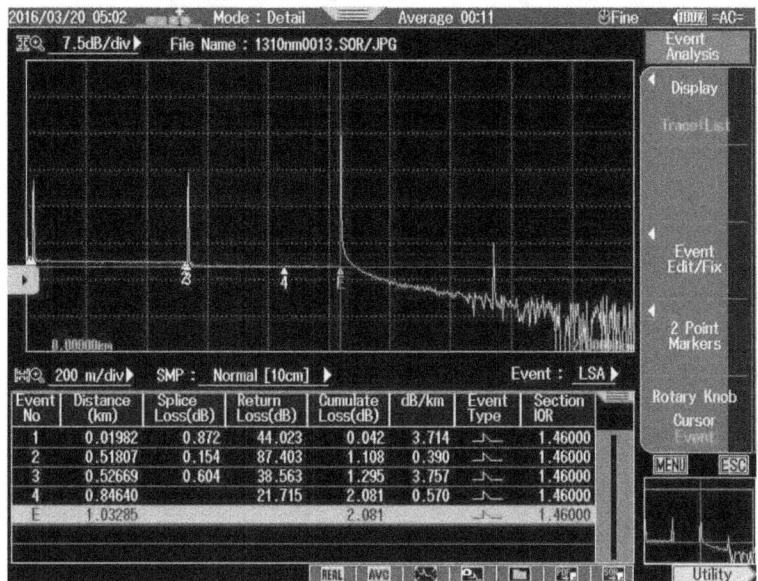

OTDR trace with event table from fiber analysis

In this section we will summarize the information in the OTDR trace and go into depth on how the OTDR measurements work in later sections.

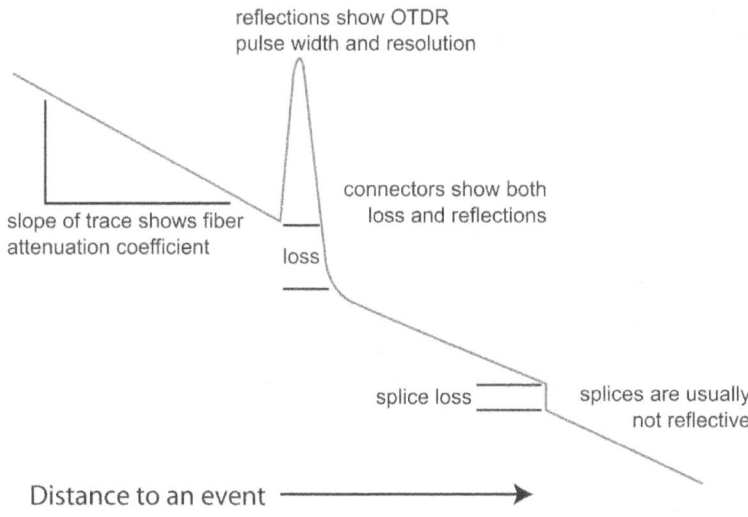

Information in an OTDR trace

Fiber Length
The OTDR measures length by calculating the time the pulse takes to travel down and back in the fiber under test divided by two to convert to a one way

distance and then using the speed of light in glass to convert time to distance. Distance = speed X time. The accuracy of this measurement depends on knowing the speed of light in the fiber accurately, which is a specification of the fiber provided by the manufacturer. The important thing to remember about this measurement is that it is the length of the fiber, not the cable. Typical OSP loose tube cables have fiber lengths that are 1-2% longer than the cable itself. ADSS (all dielectric self-supporting) aerial cables may have an even larger difference due to their construction.

Fiber Attenuation

The slope of the fiber trace shows the attenuation coefficient of the fiber and is calibrated in dB/km by the OTDR. The slope is determined by the loss in dB over the fiber length measured in km. In order to measure fiber attenuation, you need a fairly long length of fiber with no events along its length and no distortions on the ends from the OTDR resolution or overloading due to large reflections. If the fiber looks nonlinear at either end, especially near a reflective event like a connector, avoid that section when measuring attenuation.

Loss

The trace display of loss from connections, splices, tight bends or passive devices are called "events" in OTDR jargon. The OTDR can measure the loss of the joint from the decrease in optical power at the location of the event. Connectors and mechanical splices may also show a reflective peak. The height of that peak will indicate the amount of reflection at the event, unless it is so large that it saturates the OTDR receiver, losing any relevant data. Then the peak will have a flat top and usually a tail on the far end, indicating the receiver was overloaded. High reflectance events can also show "ghosts" which will be explained below.

The loss measured by the OTDR is not the exact loss of the event unless the backscatter of the fiber on both sides is identical as would happen in splicing a broken fiber. If the event is a joint between two dissimilar fibers, the loss measured will be a combined measurement of the loss of the joint and the difference in backscatter coefficients of the fibers. That difference may be large enough to cause the fiber to not show a loss, but a gain. "Gainers" are quite common with singlemode fiber tests and will be explained below.

Sometimes, the loss of a good fusion splice will be too small to be detected by the OTDR. That's good for the cabling system but can be confusing to the operator. It is extremely helpful to know the lengths of all fibers in the network, so you know where to look for events. That helps prevent confusion when unusual events show like ghosts show up on the trace, something we'll explain below.

Kinks or tight bends in a fiber causes loss that can look like a splice, a loss with no reflectance. This can be confusing especially if the kink or bend is near an actual splice. But a characteristic of optical fiber allows distinguishing a bending loss from a splice. Fiber bending losses are sensitive to wavelength, with longer wavelengths being much more sensitive to bending losses. If you take traces at two wavelengths and the loss is higher at the longer wavelength, the event is probably a bending loss not a splice.

Reflective pulses can affect the resolution of the OTDR. You cannot see two events closer than is allowed by the test pulse width. Generally longer pulse widths are used to be able to see farther along the cable plant and narrower pulses are used when high resolution is needed, although it limits the distance the OTDR can see. This topic will be discussed in the section on setting up the OTDR for measurements.

Making Measurements With The OTDR

The OTDR uses some unique ways to make measurements, based on backscatter and reflectance data. It depends on the proper additional equipment, including launch and receive test cables. It makes measurements using data from the trace chosen either manually by an operator or by the OTDR software. We cover OTDR setup later because it helps to first understand how the measurements are made to see how setup parameters affect those measurements.

Launch And Receive Reference Test Cables
To connect the OTDR to the cable to be tested, you need a launch cable and usually a receive cable on the far end. The launch and receive cables have several purposes and that affects the proper choice of cables.

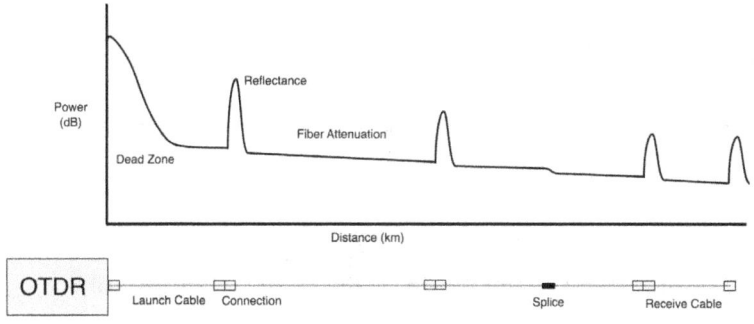

OTDR connected to cable under test with launch and receive cables

The launch cable needs to be long enough to allow the return trace to settle down after it is overloaded by the test pulse from the OTDR (creating what is called a "dead zone") and provides a connection to the cable under test to allow testing the connector on the cable. For testing short multimode cables with a high resolution OTDR the launch cable may only need to be 20 meters long, although 100 meter cables will work on all multimode systems which are generally under 2 km. Singlemode launch cables for long distance testing need to be at least 1 km long to allow for the dead zone due to long test pulses. High resolution OTDRs may use shorter cables, around 100m, when testing PON systems or premises cables.

To test the connector on the far end of the cable under test, you need to attach a receive cable to it. The receive cable is also useful in other measurement functions such as bi-directional testing. It is recommended that launch and receive cables be the same length and if possible made from the same batch of fiber which will assist in making bi-directional measurements. Bi-directional measurements will be discussed in detail in the section on measurement accuracy.

The test cables must first be chosen to be the same type of fiber as the fiber in the cable being tested. For multimode fiber, OM1 fiber should be tested with OM1 fiber. OM2, OM3, OM4 and OM5 fiber can be tested with any of the fibers since they are all 50/125 fibers and the differences are not significant. It is recommended that bend-insensitive fibers not be used as they can create measurement errors that will be discussed in the OTDR accuracy section below.

Singlemode fiber test cables should also be chosen to be the same type of fiber as the fiber in the cable under test. Different types of singlemode fibers have different mode field diameters that cause differences in backscatter levels that will cause errors at connections to the fiber under test.

It is highly recommended that launch and receive cables be the same length. This is helpful in allowing the cables to be swapped as needed. In addition, when making bidirectional OTDR measurements, you do not disconnect the launch cables but only move the OTDR itself as will be explained below. Having the cables the same length will simplify the OTDR analysis of bidirectional testing.

The connector on the OTDR should be protected. Damage to this connector can be costly and time-consuming to fix. For singlemode OTDRs, ensure you have the right connector type as many SM OTDRs use APC connectors to prevent reflectance at this connector adding to the dead zone. Trying to connect PC connectors to APC connectors can damage both connectors.

APC connectors should be color-coded green and PC connectors should be color-coded blue. Experienced users recommend using a short cable connected to the OTDR at all times and connecting the launch cable to it. Again, use of APC connectors here is recommended.

Using The OTDR Markers To Analyze Traces
OTDRs allow setting "markers" on the trace to make measurements. Newer OTDRs with color displays have different colored markers. The two primary markers set two points for making measurements in between the markers. Secondary markers are used to place measurement tools on sections of the trace for other measurements.

Whenever using markers, it is very important to ensure the marker is on a linear section of the trace that is not too close to an event. Events will have a certain width due to the width and shape of the test pulse, so placing markers too close to the event can cause errors in the measurement.

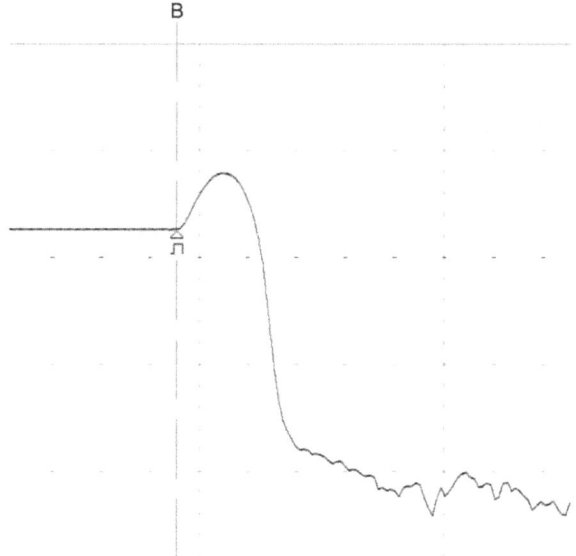

Proper placement of a marker at a fiber end, just before the reflectance peak

Fiber Length And Attenuation by The Two Point Method.
The OTDR measures distance and loss between the two markers. This can be used for measuring loss of a length of fiber, where the OTDR will measure the distance between the two markers. The OTDR will also measure the dB difference on the trace between the two markers and calculate the attenuation coefficient of the fiber.
One of the more basic uses of this test is inspecting bare fiber on a reel to

determine the length of the cable and if it has been damaged during transit.

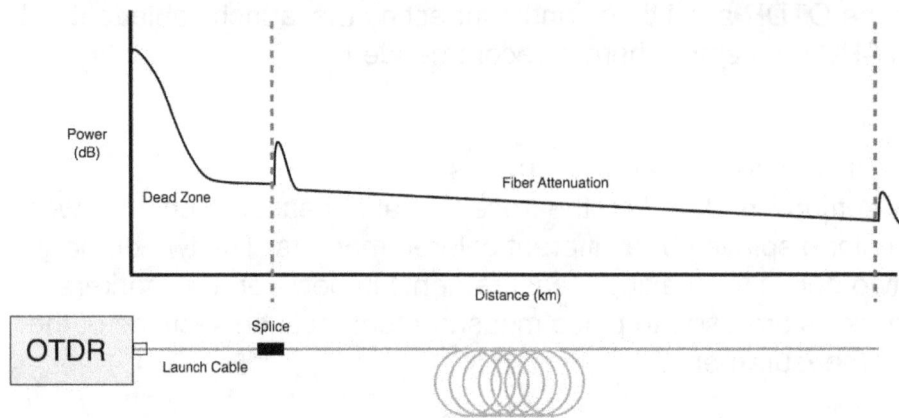

OTDR testing of bare fiber

To measure the length and attenuation of the fiber, we place the markers on either end of the section of fiber we wish to measure. The OTDR will calculate the distance difference between the two markers and give the distance, usually in km but some users work in English units like kilofeet (1000 feet.)

The OTDR also reads the difference between the power levels of the two points where the markers cross the trace and calculate the loss, or difference in the two power levels in dB. Finally, it will calculate the attenuation coefficient of the fiber by dividing loss by distance and present the result in dB/km, the normal units for attenuation.

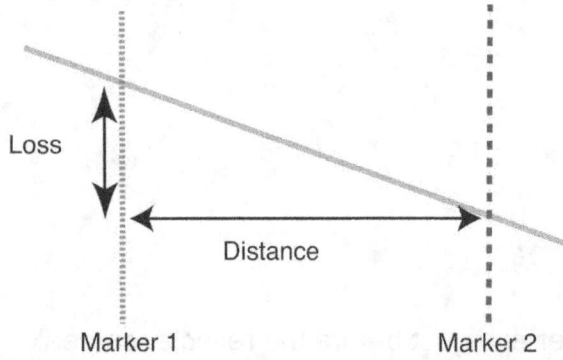

Two point measurement on an OTDR trace

In order to get a good measurement of the attenuation coefficient, it is necessary to find a relatively long section of fiber to give a good baseline for the measurement. Short distances will mean small amounts of length and loss, and the uncertainty of the measurement will be higher than if the distance is longer. It is also advisable to stay away from events on either end of the fiber section like splices or connectors, as the OTDR may have some settling time after these events, especially if they are reflective, causing the trace to have nonlinearities caused by the instrument itself.

Fiber Attenuation by Least Squares Approximation Method (LSA)
Sometimes the fiber being measured is noisy, perhaps because it is at the end of a long cable run, or has some nonlinearities which might be caused by stress on the cable during installation. OTDRs have a method of test called "least squares approximation" or LSA. Using LSA, the OTDR measures distance and loss between the two markers by calculating the statistical best fit straight line between the two points mathematically using the "least squares" method.

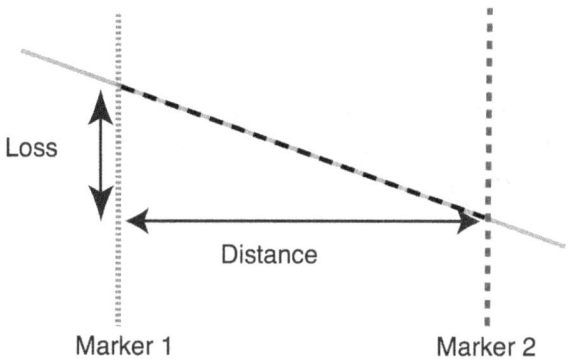

Least squares measurement (LSA) of fiber attenuation (dashed line)

When the markers are selecting the noisy part of the fiber trace, the least squares approximation (LSA) tool can be applied to calculate the dB loss between the cursors. Look closely and you will see a thick dashed line between the markers, indicating the best fit to the trace, averaging all the noise and reducing its effect on the measurement.

Testing Cables With Splices And Connections
It's more common to use the OTDR to test cables that have been installed with splices and terminations. At the same time, the traces can be inspected for stress loss caused by improper installation.

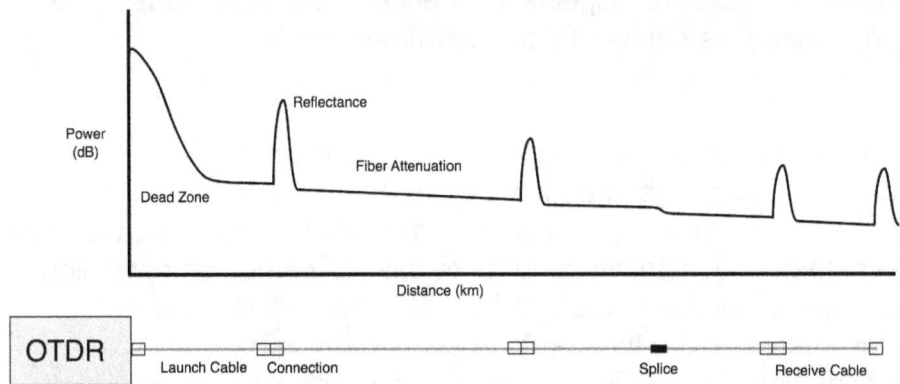

OTDR setup and trace for cable testing

Splice And Connector Loss by The Two Point Method
In measuring splice or connector loss by the 2 point method, the OTDR measures distance to the event at the first marker and loss at an event - a connector or splice - between the two markers.

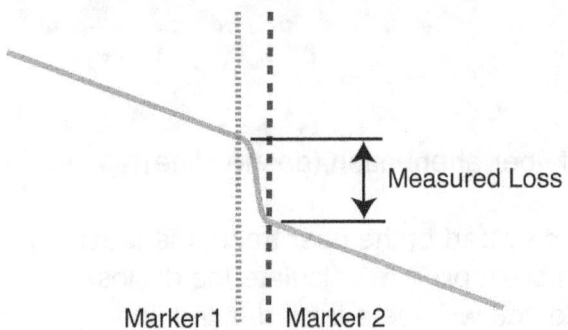

Two point loss measurement of splice

To measure splice loss, move the two markers close to the splice to be measured, having each about the same distance from the center of the splice. The splice won't look as neat as this, with the instrument resolution and noise making the trace less sharp looking in real traces The OTDR will calculate the dB loss between the two markers, giving you a loss reading in dB.

Measurements of connector loss or mechanical splices with some reflectance will look very similar, except you will see a peak at the connector, caused by the reflectance of the connector. A connector event will usually also look wider due to the reflectance of the joint and require the markers to be spaced further apart.

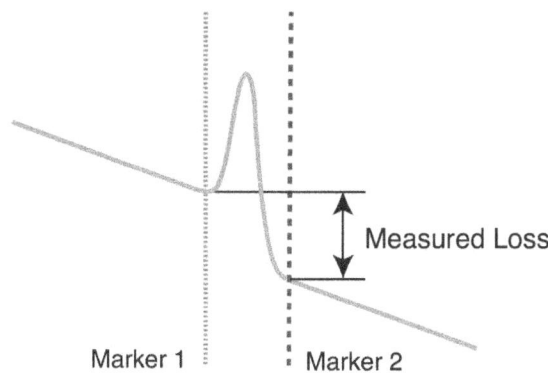

Two point loss measurement of connector

The distance between the markers is a source of error in the measurement since the actual even is only microns wide, but the markers have to be spaced tens or hundreds of meters apart due to the width of the event on the trace. That width is a function of the test pulse width and will be 2 to 3 times the width of the test pulse.

When making a 2 point loss measurement, the length of fiber between the two markers is included in the measurement, so the loss of that fiber is added to the loss of the actual event. With multimode fiber, the distance between the markers is often in the range of 10 to 20 meters and with a fiber attenuation of 3 dB/km means that 0.03 to 0.06 dB loss is added to the actual loss of the connection. On long links of singlemode fiber, the distance can be 100 to 200 meters. With a fiber attenuation of 0.4 dB/km at 1300nm, that would add 0.04 to 0.08 dB to a splice loss. That amount of error is significant, especially with a fusion splice, so a variation of the LSA test method can be used.

Splice Or Connector Loss by Least Squares Approximation (LSA)
When measuring connector or splice loss using "Least Squares Approximation" (LSA,) the OTDR uses the fiber traces on each side of the event to determine the loss of the joint. Most OTDRs offer this feature for splice and connector loss. The instrument will calculate the loss by extrapolating the fiber traces on both sides of the event and calculating offset of the fiber traces to determine the loss.

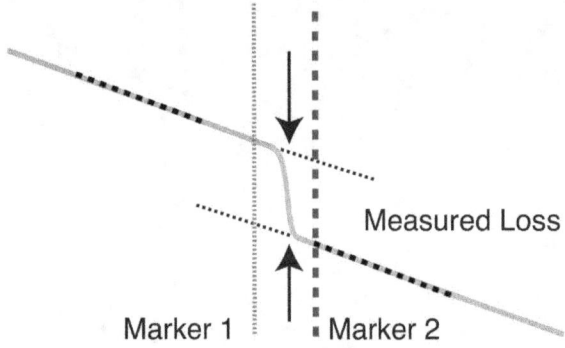

LSA analysis of splice loss

Setting LSA requires setting several markers - one on the approximate location of the event, two smaller markers on the fiber before the event and two other smaller markers after the event that define the fiber segments used for least-squares analysis. These segments should be long enough to allow good measurement but not so long as to approach other events.

By extrapolating the fibers to the location of the joint (Marker 1) and measuring the offset at that point, the OTDR calculates the loss of the joint minus any contributions from the loss of the fibers. This is helpful with connection loss because of the wider nature of the event. Therefore this measurement is less prone to error than a simple two point method.

LSA is particularly useful on connectors with high reflectance that may cause problems with the OTDR trace. The reflectance can cause the OTDR to overshoot on the recovery from the high peak producing a trace like the one below. In order to avoid the nonlinear recovery from the high reflectance peak, it is necessary to place the marker far down the trace to a point where the trace has fully recovered.

The large distance between the markers means the attenuation of a large distance of fiber is included in this measurement. This can be a big problem when testing premises multimode cables with an OTDR where the reflectance can be high and the fiber attenuation is high at 850 nm. By using the LSA method, these problems can be minimized.

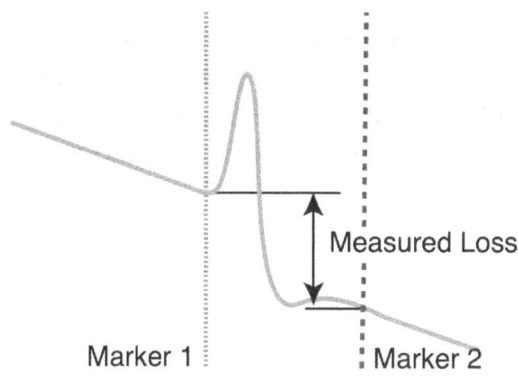

High reflectance connection loss by the two point method

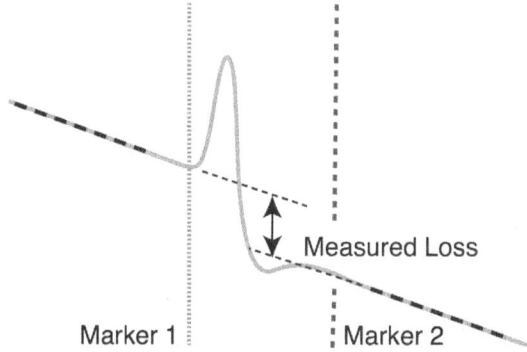

High reflectance connection loss by the LSA method

The difference in measured loss using the two methods is apparent. The two point method introduces significant measurement error due to the fiber included in the region between the markers or in case the markers are placed on nonlinear parts of the traces.

Note: The measurement of loss at a joint between two fibers has an inherent error if the backscatter coefficient of the fibers is different. We discuss that in detail below in the section Directional Fiber Joint Loss.

Reflectance Measurement By OTDR

The OTDR records the amount of light that's returned from both backscatter in the fiber and reflected from a reflective event such as a connector or mechanical splice. The amount of light reflected at a connection is determined by the differences in the index of refraction of the two fibers joined, a function of the composition of the glass in the fiber and any air in the gap between the fibers. Most fusion splices will have minimal if any reflectance, but most PC (physical contact) connectors and mechanical splices will have some reflection.

APC (angled physical contact) connectors with an angled PC interface have a reflectance of around -60 dB, and probably will not show any much reflectance on an OTDR trace. It may be necessary to refer to cable plant documentation to determine if a low reflectance event seen in a trace is a fusion splice or an APC connector.

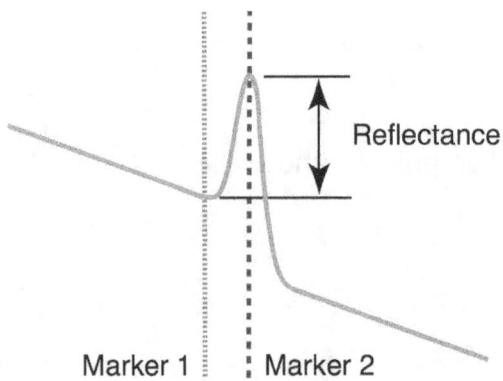

Reflectance measurement by an OTDR

By choosing the reflectance measurement mode on the OTDR and putting the right cursor on the peak of the reflection and the left cursor just to the left of the reflection at the beginning of the peak, the OTDR will calculate the reflectance of that event.

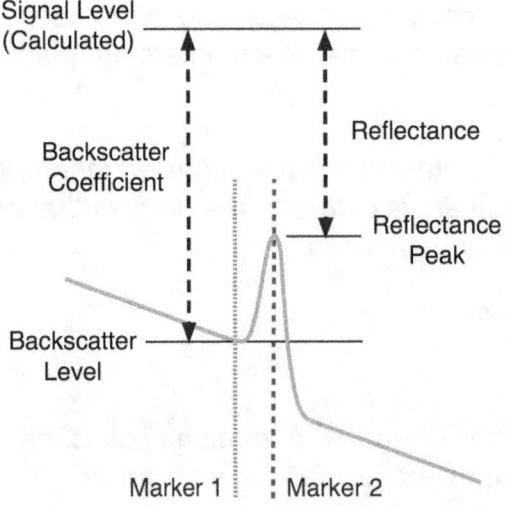

OTDR calculation of event reflectance

The OTDR calculation of reflectance is a complicated calculation involving the baseline noise of the OTDR, the backscatter level of the fiber and power at the reflected peak as shown in the diagram. Since reflectance is defined as a fraction of the amount of power in the test signal, it is necessary to first calculate the test power at that point in the trace – the "0 dB" reference power - and then calculate reflectance. Since the test power cannot be determined directly, it is implied from the backscatter level of the fiber, based on the typical backscatter coefficient of the fiber being tested.

Typical backscatter levels for optical fibers

Wavelength (nm)	Fiber Backscatter Level (1 ns test pulse)
850 (MM)	-67 to -70 dB
1300 (MM)	-74 to -76 dB
1310 (SM)	-77 to -80 dB
1550 (SM)	-81 to -82 dB
1625	-82 to -3 dB

The OTDR then measures the backscatter level just before the reflectance peak and uses the test pulse width and the backscatter table to imply test power, compares that to the power level at the reflectance peak and calculates the reflectance. The indirect way this is calculated makes the uncertainty of this method high, but provides a simpler method to measure reflectance than the OCWR method described in Chapter 11.

If the reflectance peak is large and the top of the reflectance peak is flat, it indicates the reflectance signal has probably saturated the OTDR receiver. A saturated peak cannot be used for reflectance measurement since the actual peak height is masked by the signal exceeding the dynamic range of the OTDR receiver.

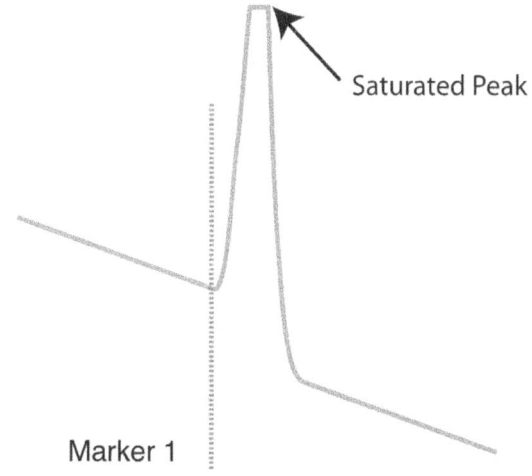

Saturated Peak

Marker 1

OTDR trace of saturated reflectance peak

Like all reflectance measurements, the OTDR has a fairly high measurement uncertainty. This is caused in part by the difficulty of positioning the cursors properly, especially the right cursor on the peak of the reflectance event, the difficulty of establishing the levels of the baseline backscatter as well as the large dynamic range often encountered. While the uncertainty of reflectance measurements is very high, ±several dB, the OTDR technique does have the advantage of showing where reflective events are located so they can be corrected if necessary.

Optical Return Loss
Optical return loss is related to reflectance. Beside reflectance from individual events, a fiber will have backscatter that the OTDR is using for making measurements. The combined amount of light from reflectance and backscatter is called optical return loss (ORL). ORL may be a factor in system total noise that affects transmission systems, so OTDRs often can calculate the ORL from the trace. It may be necessary to specify the reporting of ORL but there is no special test setup to measure it.

Cable Plant Loss Testing
The OTDR can be used for testing the loss of an entire cable plant. While this is not a replacement for the standardized insertion loss test, there are times it can be very useful. It is necessary to use both a launch reference cable and a receive reference cable to be able to include the connectors on each end of the cable under test.

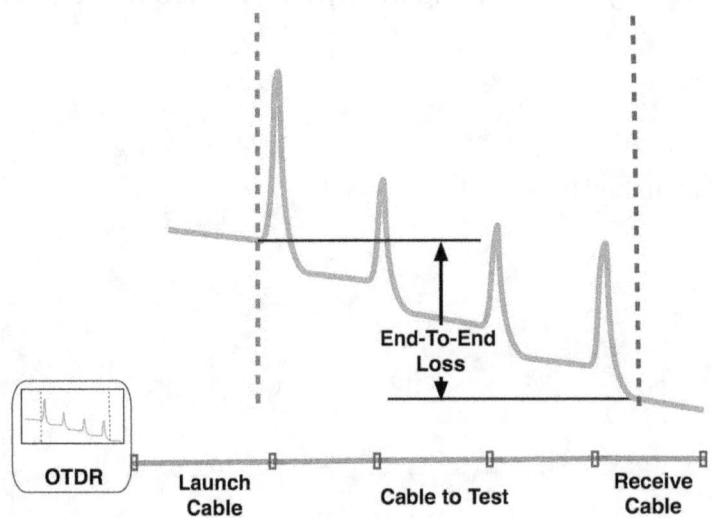

OTDR test for cable plant loss

The problem with this test is the same as with any two point test - the end marker includes some length of fiber beyond the final connection which adds to the total measured loss. One international standard (IEC 61280-4-1) has a description of a method using LSA but it is not the usual event LSA since it requires that the markers define the beginning and end of the cable plant being measured as shown below.

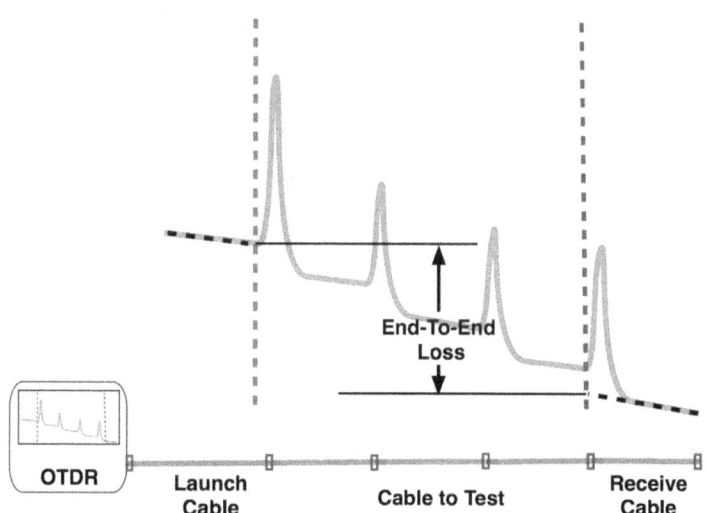

LSA method for cable plant loss

This method does not seem to be implemented in many OTDRs yet but since it is included in an international standard, it is likely to be used in the future.

Out Of Band Testing
There are two meanings to "out of band" testing with OTDRs. The first meaning is to test at 1625-1650 nm when installing fiber optic cables, above the bands used for communications where the fiber sensitivity to bending and stress losses are higher. This makes finding points of stress or losses caused by bends easier to find. Sometimes the term is also used for testing at both 1310 nm or 1550 nm when the fiber is being installed for use at only one of the wavelengths. Testing fiber intended for 1550 nm or DWDM use at 1310 nm, for example, provides a comparison for tests at longer wavelengths and helps identify stress losses since they will be minimized at the shorter wavelength.

The second meaning is to test fibers for maintenance or troubleshooting while operational (live), generally using those same wavelengths but sometimes using wavelengths not being used for transmission. Live fiber testing requires using wavelength division multiplexers (WDMs) that isolate the operational wavelengths from the testing wavelengths to prevent disrupting communications. Provision is made for this kind of testing in many modern transmission systems. DWDM systems generally use 1650 nm for in service testing for the additional isolation from transmission wavelengths.

OTDR Measurement Errors

The measurement errors caused by marker placement are but one type of errors with OTDRs. As we will see later in this chapter, the OTDR is basing this measurement on the backscatter levels of the fibers. If the backscatter levels are equal as it would be with a single fiber that has been broken and spliced back together, the loss measured would be close to the actual value. But if the fibers are different and have different backscatter coefficients, the difference would lead to an error in the measurement of the loss. We will discuss this later in the chapter.

Comparing Traces

OTDRs have the ability to compare two or more traces on the same display. All OTDRs offer this feature, where you can copy one trace and paste it on another to compare them. There are a number of reasons that you may want to compare traces.

If you take data on the same fiber at two wavelengths, you can compare the two and look for differences. In particular, this comparison can help find stress-induced losses in a cable caused by installation problems. With all fibers, longer wavelengths are more sensitive to stress losses. Traces of multimode fiber at 850 and 1300 nm will show stress loss at 1300 nm when compared to 850 nm traces, unless the fiber is bend-insensitive fiber that shows virtually no stress loss. For singlemode fiber, traces at 1550 nm show much higher stress loss than at 1310 nm and some OTDRs also offer 1625 nm sources, a wavelength where stress losses are much higher.

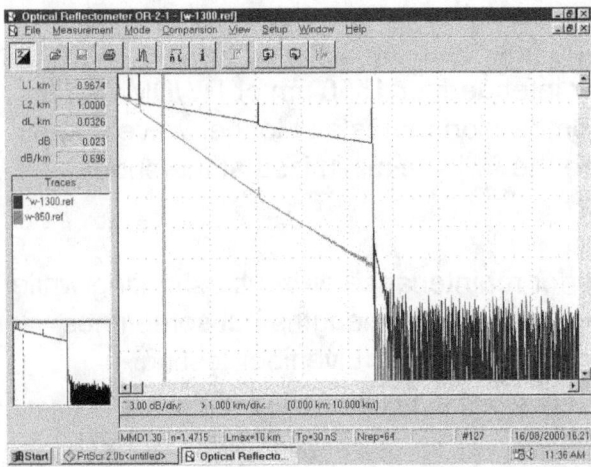

OTDR comparison of two multimode fiber traces at 850 and 1300 nm

These two traces above are taken from the same multimode fiber cable plant at different test wavelengths. The major difference in the slope of the traces displays the different attenuation coefficient of the fiber. The top trace shows the cable at 1300 nm with an attenuation coefficient of about 1 dB/km. The bottom trace shows the same cable measured at 850 nm where the attenuation coefficient is about 3 dB/km. There is also a noticeable difference in the reflectance at the connection. Seeing a variation in reflectance due to a wavelength difference is not unusual.

Comparing two or more traces also allows comparing a fiber to other fibers in the same cable. In theory, all the fibers should be similar, so comparing them makes it easier to spot problems. If two fibers have markedly different overall losses, comparing them should allow the operator to more easily spot the cause of the differences, and the OTDR will provide the location of the area of concern.

One of the more important uses of comparison is when troubleshooting for restoration. If a network goes down because of a fiber problem, having prior data to compare to current data makes it easier to spot problems for repair. A copy of all data from the installation should be stored with restoration documentation for reference if a problem arises.

When setting up the OTDR for a test, it may also be useful to compare traces taken at different test pulse widths to decide which setting gives the best compromise between noise reduction and event resolution.

OTDR Setup Parameters For Best Test Results

OTDRs have a number of setup parameters that affect the measurements obtained. Most of these relate to the specific fiber being tested, the wavelength of the test, the length of the fiber and how one optimizes the OTDR setup to get the most useful data on the fiber.

Most OTDRs offer an automatic test setup that chooses these test parameters for the user. Considering the power of the processors used to manage the OTDR and analyze the data, this function should be able to choose parameters as well as a human operator. However, based on experience with various OTDRs, discussing this function with OTDR engineers and examining data from users, it appears that most OTDR manufacturers compromise this function to reduce the time necessary to take a trace. One should not blindly trust this function as the compromises mean that the OTDR will often choose the wrong parameters.

Experienced OTDR users recommend not using auto test function for testing high fiber count cables, although it would seem that would be an ideal application since all fibers are the same. However, auto test takes longer to make a test than manual setups and if there are differences in fibers, even subtle ones, it can cause the auto test function to choose different setup parameters making it harder to compare fibers in the cable. One example is a bad connector connecting to the launch cable may cause significant power loss and the choice of different pulse widths or averaging. Another situation where auto test may switch parameters is a spliced "tree and branch" network where there may be two or more sets of optimal settings.

If auto test is used, it can be used on the first fiber and if it produces acceptable results, switch to manual mode and use those same parameters. OTDRs may save the auto test parameter when you switch to manual.

Below are some of the important OTDR test parameters that need choosing for a specific fiber test.

Fiber Type And Application
The first parameter to choose before making a measurement is to choose the proper fiber type and application. Most OTDRs have plug-in modules that are specific to one fiber type, either multimode or singlemode. Some OTDRs offer multiple modules for singlemode, which represents the largest number of applications for OTDRs. These singlemode modules allow specialization for average networks like in a metropolitan area, very long distance networks or short networks like FTTH PONs or premises cabling.

Wavelength
Just like for insertion loss testing, one needs to choose a source at the proper wavelength, typically the wavelength the communications system operating over the fiber will be using. OTDRs will typically offer multimode testing at 850 or 1300 nm and singlemode testing at 1310 and 1550 nm, although some also offer 1625 or even 1490 nm. 1625 nm is useful when looking for stress losses and 1490 is used for PON networks and fiber amplifiers. Some OTDRs can also measure at multiple wavelengths for spectral attenuation and chromatic dispersion (CD) testing used in fiber characterization. Fiber characterization will be discussed in Chapter 8.

Range
The range of the OTDR is the total distance the OTDR analyzes. It is best to start with the range set at two times the expected length of the cable under test. Some OTDRs will create confusing displays if the range is set too short

because the test cycle depends on the range. If the range is set too short, the instrument may send out a second test pulse before the returned signals from the previous pulse have been processed, creating a confusing display.

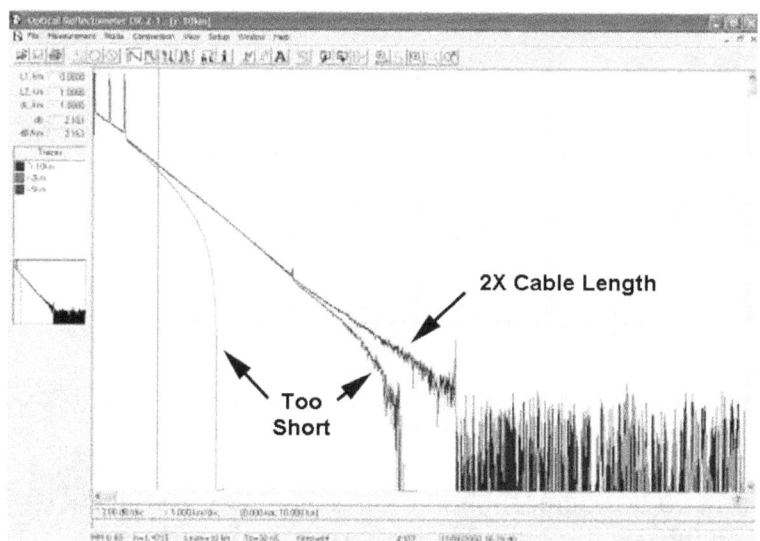

OTDR traces set at proper range and two ranges that are too short for the cable under test

Pulse Width

The OTDR will allow the choice of several pulse widths. Shorter pulses have less total light energy to excite backscatter and longer pulses have more energy and will excite more backscatter. The shorter pulses will provide a shorter dead zone and greater resolution in the trace but will limit the distance the OTDR can reach. Longer pulses allow for measurements of longer cables but result in a greater dead zone and poorer event resolution. Less resolution means that two close events may be merged into one even in the OTDR trace or low loss splices may not be visible at all.

The length of the dead zone will determine the minimum length of the launch and receive cables. For greatest flexibility in the use of the OTDR, it is recommended to choose launch and receive cables long enough to accommodate the longest test pulses expected to be used so the cables will work with all OTDR setups.

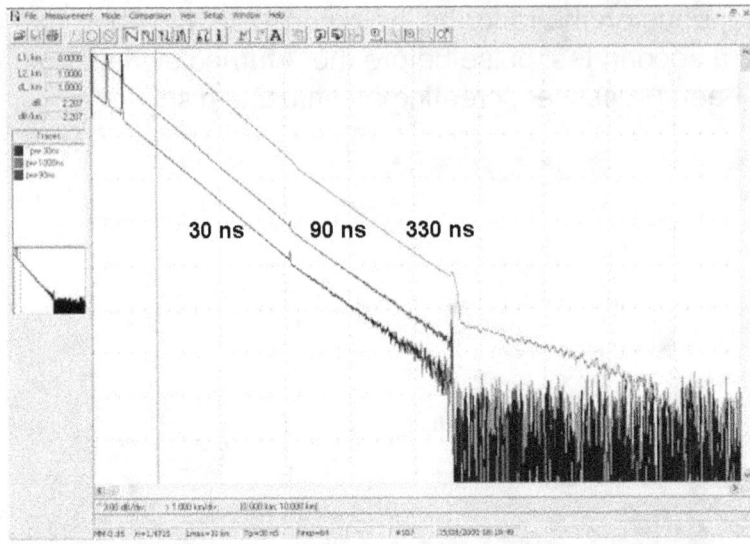

OTDR traces at 30, 90 and 330 ns showing the effect of pulse width

In the trace above you can see the difference in the backscatter from the traces where longer pulses makes the backscatter level higher and the noise at the end of the trace much lower. The significant difference in noise allows the OTDR to analyze fibers that are much longer. However the higher backscatter and longer dead zone caused by the longer pulse means the dead zone of the OTDR increases also. Note in the trace above the OTDR is overloaded in the dead zone.

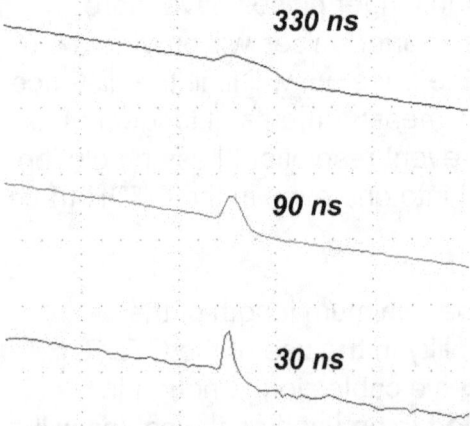

330 ns

90 ns

30 ns

Traces of an event showing the effects on resolution with 3 different pulse widths

Wider pulse widths tend to spread out events as shown above and make them harder to distinguish from the backscatter background. This can make it more difficult to locate low loss splices or connectors with little reflectance as in the trace above.

Wider test pulses will also affect the ability of the OTDR to resolve two close events. In the diagram below, you can see that a wider test pulse will merge the two splices into one event while the shorter pulse will be able to resolve them.

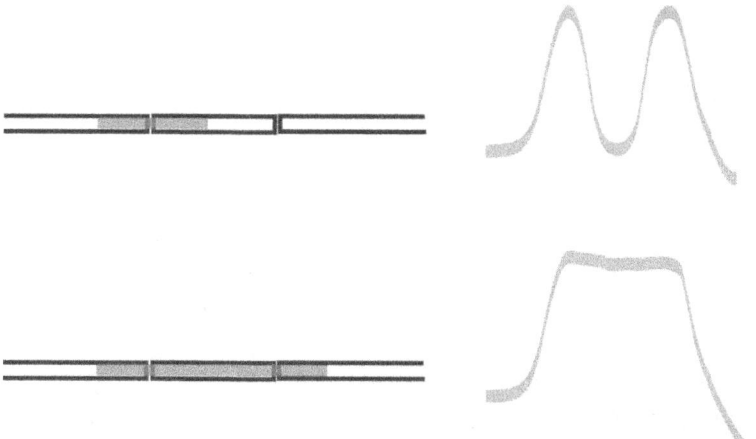

Trace resolution with different pulse widths

The test pulse width also affects OTDR reflectance measurements. The amplitude of a reflective feature on an OTDR trace appears larger when measured by shorter pulse widths as you can see in the three traces above. This affects the measured reflectance and adds a potential error to the reflectance measurement.

Guidelines: Use short pulses on short cables and long pulses on long cables. On very short cables, less than 1 km, use the shortest pulse or the next to the shortest, typically 3-10ns. For very long cables, use pulse lengths in the 500ns to several microseconds range. In the middle ranges, 20-100ns should work best. Remember you can use pulse width and averaging interactively to get the best resolution and measurement.

Averaging
Another option to get more distance and lower noise with the OTDR is to average the trace more times. The two traces below are taken of the same cable plant. The trace on the left is not averaged at all and shows the inherent noise in a single trace. The trace on the right has been averaged 1024 times. Averaging reduces the noise considerably, but averaging takes time. To get the right trace, the OTDR made 1024 tests, stored the data and averaged the results to get the smoother trace on the right.

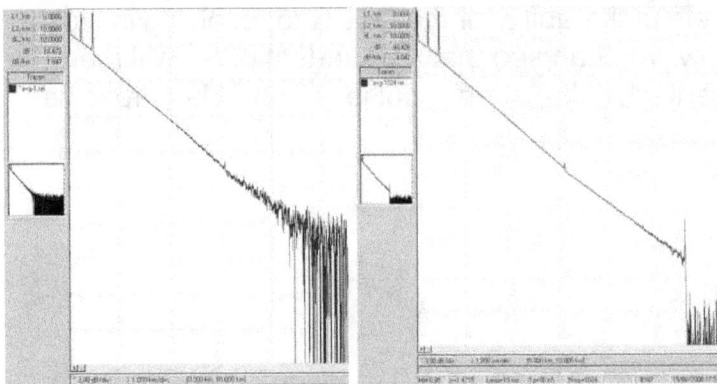

OTDR traces of multimode fiber at 850 nm with no averaging (L) and 1024 averages (R)

Choosing the right combination of pulse width and averaging requires making compromises. It is better to make measurements fast but lose the ability to resolve close or low loss events or to take longer to average traces and maintain the best resolution. The answer is to look at the fiber type, cable length and types of events and perhaps test the options on one fiber before testing all the fibers in a cable.

You will see different effects with multimode and singlemode fiber testing. Multimode fibers have higher loss because they have higher backscatter, while singlemode fiber has lower loss and backscatter, so the two factors somewhat cancel out. Multimode fibers will become more noisy on shorter cables than singlemode.

Guidelines: Shorter cables require less averaging to get reasonable noise at the end of the trace. Try 32-64 averages for short cables, 512-1024 averages for very long cables and 64-512 averages for middle lengths. Make tradeoffs with pulse width to get the best compromise.

Index of Refraction
The index of refraction (IOR or N) of the glass in the core of the fiber is used to calculate the fiber length. The IOR determines the speed of light in the fiber. The speed of light in the fiber is the speed of light in a vacuum (C, 299,792 km/s) divided by the IOR. For a popular singlemode fiber, Corning SMF-28, the IOR at 1310 nm is 1.4675 (see table below), so the speed of light in the fiber is:

Speed = C/IOR = 299,792 km/s/1.4675 = 204,287 km/s

Knowing that speed, the OTDR measures the time and converts it into distance. As you can see from the table below, the difference between most popular fibers is quite small, less than 1%. That difference is small enough to generally ignore unless you know exactly what fiber you are measuring.

Group Index of Refraction (IOR) – Single Mode Fibers

Manufacturer	IOR @ 1310nm	IOR @ 1550nm
Lucent Std. Single Mode	1.468	1.468
Lucent TrueWave and TrueWave-RS	1.471	1.470
Lucent AllWave	1.466	1.467
Corning SMF-28	1.4675	1.4681
Corning SMF-LS	1.471	1.470
Corning E-LEAF	N/A	1.469
Alcatel SMOF	1.464	1.4645

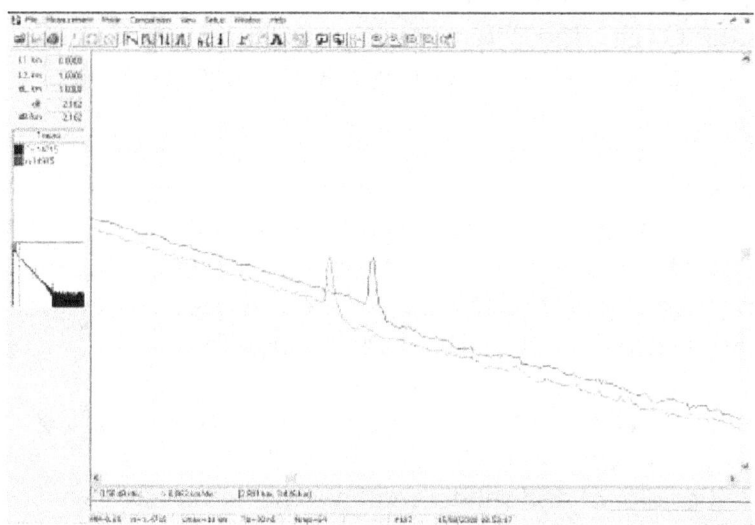

Two traces of an event in the same cable at different IOR

Since fiber optic cable has about 1% excess fiber, the actual cable length is less than the fiber by that amount. The OTDR makes its measurements on the fiber, not the cable, so one must estimate the cable length. If you have a long length of cable with distances marked on it, you can measure it with the OTDR and use the index of refraction to calibrate to the actual cable length and use that in the OTDR. If you do this, we suggest you make measurements on several fibers and average.

OTDR Measurement Uncertainty

Like all instruments, OTDRs have uncertainty in their measurements. There are several aspects to consider with OTDR measurement uncertainty: How accurate are OTDR measurements? How do they compare to insertion loss tests? How do different OTDRs compare?

We have previously covered the issue of OTDR vs. insertion loss testing. They use different methods that often do not correlate. OTDR measurement correlation between different instruments is an issue of calibration which we cover below. Let's focus on the accuracy of the OTDR measurement itself and later in Chapter 14 we will examine all these errors in depth.

The biggest cause of error in OTDR measurements is the measurement technique itself. Unlike the light source and power meter that measures loss in the same manner as a transmission system works, the OTDR uses an indirect method of measurement based on backscattered light. Some test variables like source wavelength or the quality of launch and receive cables affect OTDR measurements in a similar manner as insertion loss measurements. But with the OTDR, the difference in backscatter coefficient of two fibers at a connector or splice can cause errors in loss measurement.

Another major error in OTDR measurements is caused by the placement of the markers that determine where on the trace the measurements are made, even if you use the "least squares" mode to reduce the variations. This creates a random error that affects the precision of the measurement. If you use an "auto test" function, the placement of the markers is done by software and the reproducibility will be a function of the software algorithms.

Software is also an issue for OTDRs. Software is very much a part of OTDR measurements. It controls the setup parameters, controls the test pulse, and averages the backscatter signal and reflectance signals to create the trace. If you use the OTDR auto test function, software decides how to set parameters, where to place markers and how to make measurements.

Software, therefore, plays a big role in the accuracy of OTDR measurements. The role that software plays in the measurement is hard to evaluate. You could test the same cable plant with two different OTDRs using different software and compare the results, but many of the variables noted above also contribute to the variability of the measurements and make isolating variables difficult.

OTDR Calibration

One issue you rarely hear about is OTDR calibration. The measurements depend on the accuracy of the timebase for length measurements as well as the IOR of the particular fiber and the linearity of the input amplifier over the whole dynamic range of the OTDR which can be 40 dB or more.

OTDR calibration is not a simple task like calibrating power meters or setting the 0 dB reference for insertion loss testing. The debate over OTDR calibration has always been whether to use a "standard fiber" method of calibration that involved calibrating every OTDR to read the trace of the standard fiber identically or an electronic method of calibrating the OTDR timebase and receiver that did not involve fiber at all.

National standards laboratories considered the options for OTDR calibration in the 1980s. They first considered making a transfer standard – a "standard fiber" - a sample fiber of known index of refraction and length with splices and connectors of known loss. To use this standard, it would require exact specification of the OTDR setup for test parameters, which adds a secondary calibration process, for example calibrating the amplitude and pulse width of the test pulse. This project was never completed as it would require manufacturing many different "standard fibers" and the method was not agreeable to all interested parties.

An alternate proposal was an external source calibration based on a device that would simulate the return signal that creates the trace. That involves an instrument that would be triggered by the OTDR test pulse and would then generate an optical power declining over time to simulate the OTDR trace.

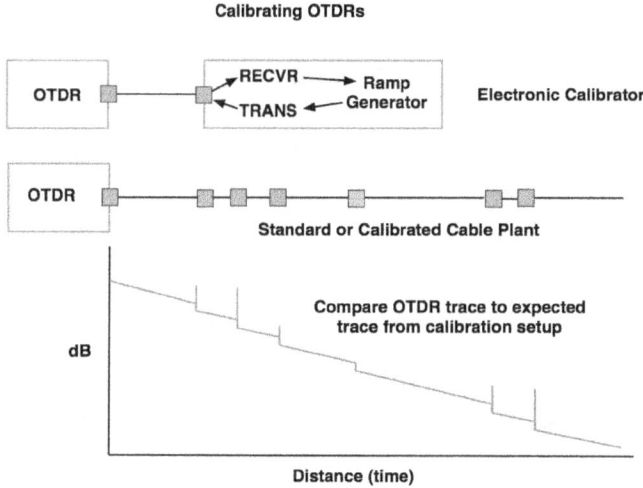

Neither method gained widespread acceptance. Standards have been written, however. IEC 61746-1 provides a standard for singlemode OTDR calibration using both methods and TIA/EIA-455-226 is a TIA adoption of the IEC document. This standard offers the two methods noted above plus an obscure third method based on fiber optic delay lines.

Since there is no readily available standard fiber or source calibrator, perhaps the best method of "calibrating" the instrument is sending it back to the manufacturer who can test the timebase and receiver linearity and confirm their performance. And, of course, they can do all the other updates, especially software, for the given model of OTDR.

Modal Conditioning In Multimode Fiber

Modal conditioning for OTDR testing of multimode fiber is a poorly understood subject. The OTDR launches a test signal with a very powerful laser that has a very restricted modal distribution. But the backscatter signal that provides the information for the OTDR to analyze basically fully fills the modes in the fiber for the return path. Therefore you have two very different analyses going on for each test – underfilled modes on the outgoing signal, overfilled modes on the return signal.

Because of this issue, mode conditioning on the outgoing signal seems to make little sense and some proposed encircled flux mode conditioners for OTDRs are not usable in a reverse direction. This is just another part of the difference in testing methods between insertion loss and OTDR testing that indicates that the two test methods are unlikely to be comparable in most test situations.

Directional Fiber Joint Loss Errors Caused By Variations In Backscatter Coefficient

When the OTDR test pulse goes through a joint in the fiber made by a splice or a connector, some light is lost in the joint. The reduced light in the test pulse will reduce the amount of light backscatter and that will be measured by the OTDR and shown as a loss in the trace at that point. That loss can be measured using the techniques described above.

The biggest source of measurement uncertainty that occurs when testing loss at a joint with an OTDR is a variation of the backscatter coefficient, the amount of light from the outgoing test pulse that is scattered back toward the OTDR by the fiber. The OTDR looks at the returning signal and calculates loss based on the amount of light it sees coming back and there is no way the OTDR can determine if the backscatter coefficient of the fibers is a constant.

The light scattered back to the OTDR for measurement is not a constant for all fibers - in fact it may be slightly different for any two fibers from the same manufacturer. Backscatter is a function of the attenuation of the fiber and the diameter of the core of multimode fiber or mode field diameter (MFD) in singlemode fiber. Scattering is the major cause of attenuation in optical fiber. Higher attenuation fiber has more attenuation because the composition of the glass in its core scatters more light. In singlemode fiber, a larger mode field diameter (MFD) results in less backscatter and inversely, a smaller MFD will have more backscatter. In multimode fiber, bend-insensitive fiber with its core/cladding design that reflects light back into the core has higher backscatter than non-BI fiber.

When the loss of a splice or connector joining two different fibers is tested in an OTDR, any difference in backscatter from each fiber will cause an error that is dependent on the direction of the test. There are three possibilities at the joint of the two fibers.

If both fibers are identical, such as when splicing a broken fiber back together, the backscattering coefficient will be the same on both sides of the joint, so the OTDR will measure the actual splice loss in each direction.

If the fiber nearer to the OTDR has higher backscatter than the one after the connection, the amount of backscattered light will go down after the joint, so the measured loss on the OTDR will include the actual loss in that direction plus a loss error caused by the lower backscatter level, making the displayed loss greater than it actually is.

If the fiber nearer the OTDR has a low backscatter level and the fiber after the joint a high backscatter fiber, the backscatter level goes up, making the measured loss less than the actual loss in that direction. If the change in backscatter level is greater than the splice loss, the joint will be a "gainer" - what looks like a gain in the fiber at the joint, not the real loss of the joint itself, a major confusion to new OTDR users.

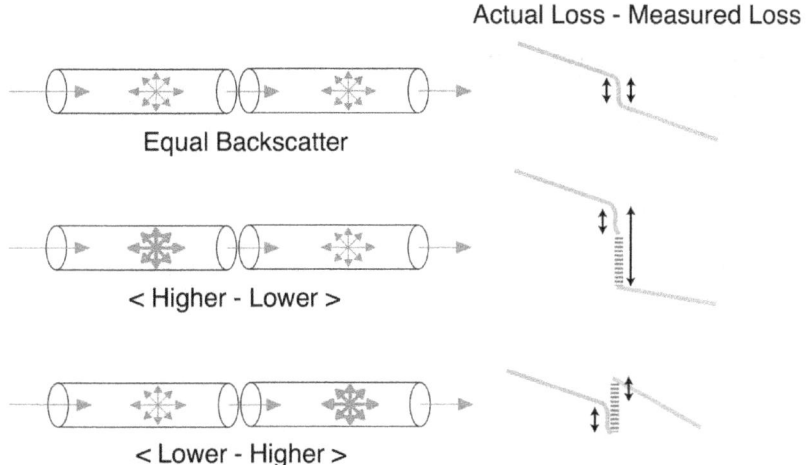

OTDR measurement of a joint between two fibers.

In the drawing of the traces, the wide dashed line in the OTDR trace of
the splice represents the difference in backscatter coefficient for the two
fibers. That amount of change can be significant. For singlemode fiber, the
tolerance in mode field diameter allowed by standards can create a maximum
difference in backscatter of almost 1 dB, although more typical production
variations will lead to variations less than half that, about 0.4 dB. Even so, the
average fusion splice is typically less than 0.1 dB, so one would expect to see
a lot of errors. And in fact, you do. Techs report that around 1/3 of all splices
will show a gain in one direction, so you know the loss is going to be in error,
showing too high on the trace in the other direction.

Here are two ways you may see gainers. First is a simple splice gain at the
event in the OTDR trace.

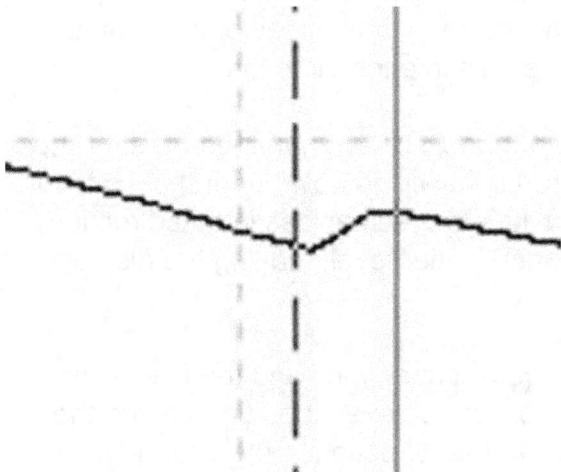

A real "gainer" - a splice 35 km away in an installed fiber link

Another way gainers show up in an OTDR trace is caused by a high
backscatter fiber spliced into a very long concatenated cable. In this case,
event 1 appears to be a fiber with smaller mode field diameter was spliced
into a link between two fibers of larger mode field diameter. At the end nearer
the OTDR, the splice shows a gainer due to higher backscatter from the
smaller MFD fiber, while at the other end it shows a larger splice loss as the
smaller MFD fiber is spliced to a larger MFD fiber. The difference is not small -
in this case it is more than 1 dB.

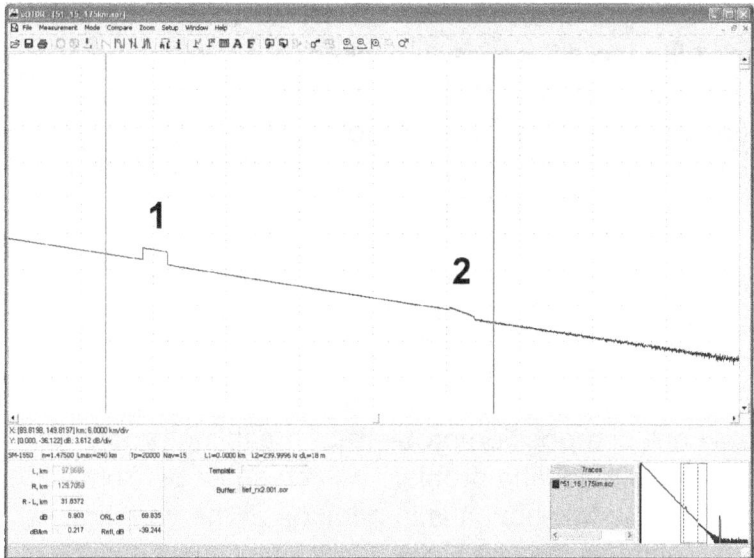

Trace of long concatenated fiber link with gainers

Event 1 shows another important issue about gainers - in the opposite direction with the opposite change in backscatter, a gainer becomes a big loser. The same difference in backscatter coefficient that causes the gainer in one direction causes a higher loss in the other direction, which can cause considerable problems in evaluating splices if you only test in one direction.

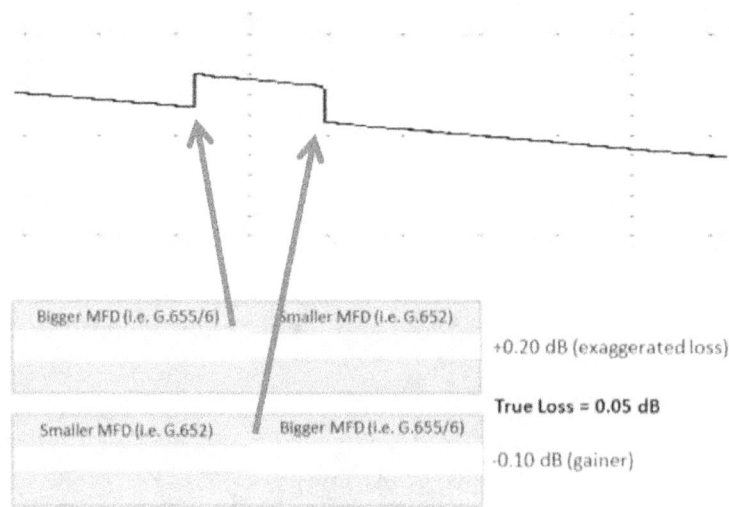

Trace of smaller MFD fiber spliced to larger MFD fibers

Here is an example of what happens when you splice a smaller MFD fiber with higher backscatter to two larger MFD fibers with smaller backscatter. When you see a trace like this, with a section of fiber elevated above other fibers (or below the other fiber traces in the opposite situation), you know it's a fiber mismatch.

In the higher loss direction, the loss shown by the OTDR will always be the actual loss in that direction plus the difference in backscatter coefficient, so even if the loss from the actual splice is very low, the measured loss can be very high. This can cause problems if you try redoing the splice trying to correct the problem, because the OTDR measurement will always be wrong and always show the loss as high no matter how good the splice.

You can see another variation with event #2. Note how the attenuation slope of the fiber is much higher in the fiber in this segment? The fiber in event 1 had an attenuation coefficient similar to the other fibers in the link, so the difference was probably just a MFD variation. Here the fiber has much higher attenuation, so it may be a core composition problem or a cabling problem that put stress on the fiber. If it is a cabling problem, which could be from too much stress on the cable in installation, it is in addition to the variation in backscatter that causes the initial gainer.

The traces below show that this issue is not limited to singlemode fiber. These traces were made as part of an experiment to see what happens when regular and bend-insensitive multimode fibers are joined. As you can see the BIMMF has much higher backscatter and creates a gainer.

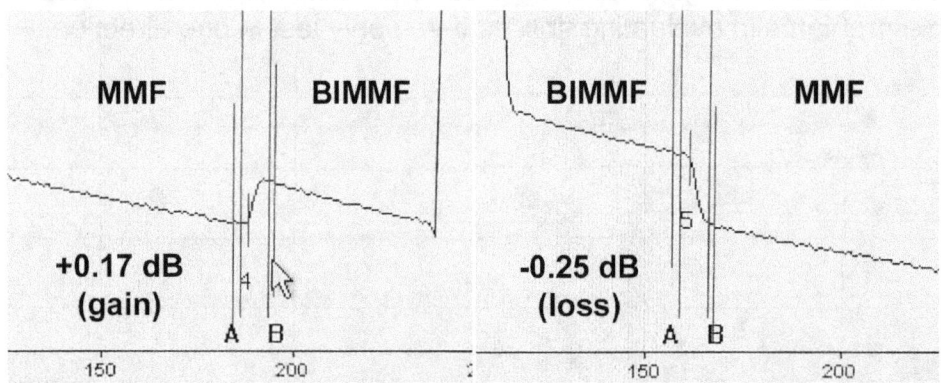

Bidirectional traces of multimode fiber spliced to bend insensitive multimode fiber

There Are Also Real Directional Loss Differences At Joints

Besides the directional loss differences caused by differences in backscatter, there are real directional differences in the loss of many joints. These may be caused by differences in the fiber designs or manufacturing variations.

Within singlemode fibers, there are regular SM fiber, large MFD fibers, dispersion shifted fibers and bend-insensitive fibers, plus manufacturing tolerances that can cause fiber differences. Some of these fibers have very different fiber index profiles. Mixing these fiber types often occurs because of the standard practice of splicing regular singlemode fiber pigtails on cables with fibers of any type.

Multimode fiber has manufacturing variations in core size among each fiber type, variations in index profiles of various bandwidth grades and bend-insensitive fiber variations.

These directional differences in loss are small but real and are very hard to measure. They require very careful lab procedures to isolate other variables, not something that can be done with field OTDR or insertion loss testing.

When you do bidirectional OTDR testing of a splice or connection, you measure the loss and backscatter effects in two directions and use some simple math (shown below) to remove the backscatter differences at the joints. While you can take out the backscatter differences with math, you end up averaging the loss from each direction, getting an average, not the actual loss.

Calculating The Average Loss From Bidirectional Data
These two traces also illustrate how one reduces the errors caused by variations in backscatter by bi-directional measurements. Note the splice loss is +0.17 dB in one direction and -0.25 dB in the other direction.

For the purpose of this analysis, we assume the actual loss is the same in both directions, which may not always be true. However this method gives an average of the losses in both directions that is adequate for most analyses.

In one direction MMF>BIMMF, we have the splice loss, L plus the additional backscatter light V, so in that direction the measured loss is:

$$Loss = L + V$$

In the opposite direction, we have the loss L increased by the variation V or:

$$Loss = L - V$$

Some simple algebra gives us:

$$L + V = + 0.17 \text{ dB}$$

$$L - V = - 0.25 \text{ dB}$$

Solving for loss,

(L+V) + (L-V) = +0.17 – 0.25 = - 0.08 dB

(L+V) + (L-V) = L+ L + V – V = 2L

2L = -0.08 or L = - 0.04 dB

So the average loss is – 0.04 dB

Similarly, solving for V,

(L+V) - (L-V) = +0.17 – (– 0.25) = - +0.17 + 0.25 = 0.42 dB

2V = 0.42 dB or V = 0.21 dB

Therefore the difference in backscatter level for the two fibers is 0.21 dB

This is the way you analyze bidirectional measurements to get the average loss.

The Magnitude Of The Directional Error Estimated From A Trace
This OTDR error source is always present, it can be practically eliminated only by taking readings both ways and averaging the measurements Many OTDRs have this programmed in their measurement routines. This is the only way to test inline splices for loss and get accurate results.

As mentioned earlier, with singlemode fiber, the tolerance in mode field diameter allowed by standards can create a maximum difference in backscatter of almost 1 dB, although more typical production variations will lead to variations less than half that, about 0.4 dB. Unfortunately, most times the tech testing the fibers has no data on the MFD of the fibers. But there may be an easier way of estimating the potential error.

The OTDR will give you data on the attenuation coefficients of the fibers on either side of the joint. That data will give an indication of potential problems. Back in the late 1980s we participated in several experiments to understand the reason behind gainers that led to understanding the cause, the variations in backscatter levels. By analyzing large amounts of data, we were able to determine the relationship with differences in MFD and loss variations. But we also had data that showed equal relationships between differences in the attenuation coefficients of the fiber and the directional variations in splice loss.

The correlation we found was that a difference in attenuation coefficient in singlemode fiber of 0.1 dB/km would lead to approximately a 0.4 dB directional difference in splice loss. By looking at the attenuation coefficient of the fibers on each side of a joint, one can estimate the directional variation and while it's only an estimate, it does give a good indication of whether one needs to make bi-directional measurements to have valid data. We have seen recent data from modern fibers that confirm our data from those earlier experiments, so this method is still useful.

Bidirectional OTDR Test Procedures

When making bidirectional OTDR measurements, do not disconnect the launch and receive cables but only move the OTDR itself to the other end. The reason is you want to test the connections at each end of the cable in both directions so you do not want to disturb those connections. That will also change the reference test cable's characteristic backscatter coefficient at that connection that was part of the original test in the opposite direction.

Having the launch and receive reference cables the same length will simplify the OTDR analysis of bidirectional testing. Using a short cable on the OTDR to connect to the reference cables will prevent damage to the OTDR connector due to frequent mating with the reference cables.

Most OTDRs have a bidirectional test function in the analysis software. Follow the OTDR instructions on how to save the traces in each direction and open them in the display for the OTDR to calculate the bidirectional data. You may also manually measure the joint loss in each direction and average.

Length Measurement Accuracy

Many fiber optic measurements depend on length. Measuring the attenuation coefficient of fiber is one of them, involving both length and loss. OTDRs are often used to measure cable length and distance to a fault, which is especially important when finding the location of damage to a cable, often caused by visible accidents like dig-ups, but sometimes by hidden damage like rodent damage, cuts from directional boring, lightning strikes, or similar problems. In these cases, knowing the cable length is more important than the fiber length.

Most OSP loose tube cables have 1-2% excess fiber (less on ribbon cables) to prevent fiber stress under cable tension encountered in pulling or aerial installation. Some manufacturers of cable can provide the correct index of refraction (IOR) to use for that cable for your testing. If you do not know IOR or the ratio of excess fiber, you can estimate it or, if you have a long spool of cable, calibrate it.

To calibrate the correction factor for a given cable, measure several fibers on the spool of cable with the OTDR. Then look at the cable jacket markings to get the actual length of the cable from the printed markings at each end of the cable. Compare the actual cable length to the OTDR measured length. Use the OTDR's setup feature to set the index of refraction to the value that makes the OTDR read the same as the marked length of the cable.

If you do calibrate the IOR for a cable, record that data with the test data. In the future you can use that IOR to have a more accurate length reading. That can be very useful in situations where cables have been damaged.

If you are troubleshooting a break in a long cable run but don't know the correction factor for fiber vs. cable length, you may still be able to get a calibration. If you have the data from the original design and testing, you may have the actual length of the cable plant. With that you can calculate the point of the break very closely. Here is an example:

Consider a 10km cable with a break around 6km from one end. From one end, the OTDR says the distance to the break is 6500m and from the other end it says it's 4000m. That adds up to 10,50 m, which we know is too long. If the actual cable length is 10,000m, the correction factor is:

Actual length/measured length = 10000/10500 = 0.952 = correction factor

Thus our 6500m measurement is actually 6500X0.952 = 6190m and from the other end it's 3810m.

You can also use this method on a cable where you do not have a length. Simply drive along the cable route or use a measuring wheel to measure the length, then use this calibration method.

Another source of error when measuring length is the placement of markers. The placement of a marker can be affected by the shape of the trace where the marker is being placed, which can be a reflectance pulse if it is at a connection or an open connector, a loss event without reflectance such as a fusion splice, or a fiber end which may or may not have reflectance.

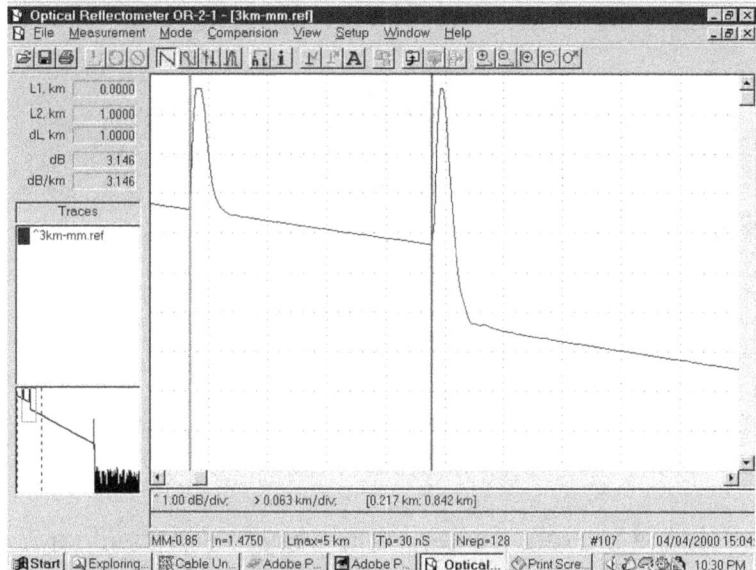

Manual placement of markers

When placing markers manually, it is important to be consistent. The markers should be placed just before the event. If the event is a connection and has a reflectance peak, the marker can easily be placed just at the point the peak rises from the backscatter level of the trace. If it is at a splice, place the marker just before where the splice loss drops from the backscatter level.

When measuring the length of a fiber or the distance to a break, finding the end of the fiber can be a problem. If the end of the fiber is cleaved or has a polished connector which has a high reflectance, it is easy to find from the strong reflectance peak.

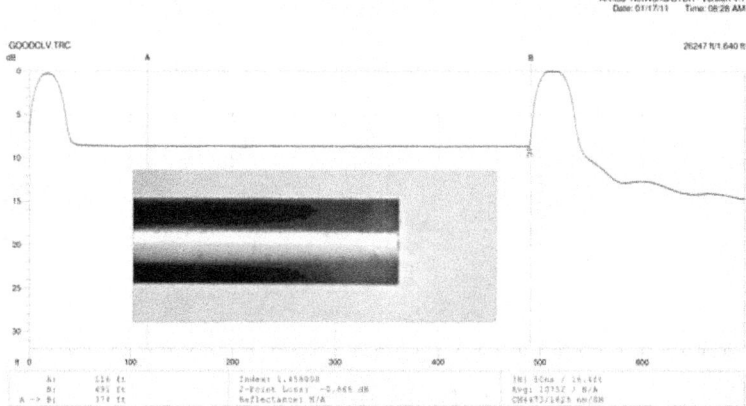

OTDR trace of cleaved fiber end

However a break will have a much smaller reflectance peak if any peak at all, making the end of the fiber harder to locate.

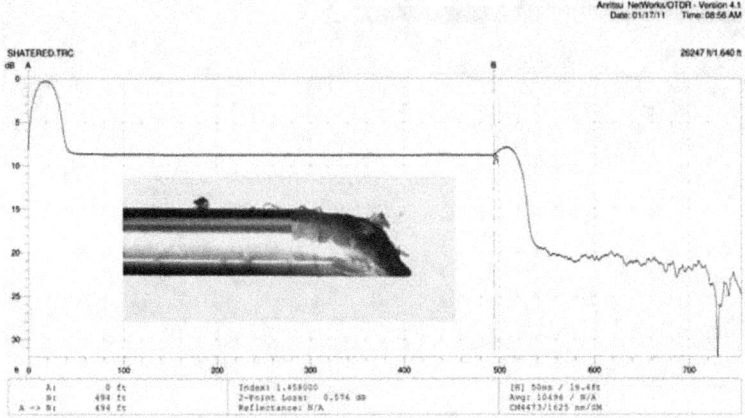

OTDR trace of broken fiber end

If the fiber is short and the end is broken, it's generally easy to see the fiber end and leave nothing but noise. On longer fibers that are near the limit of the OTDR, the trace itself may be noisy and the end hard to find. In that case, use a longer pulse width and/or more averaging to reduce the noise on the fiber so you can see the transition from backscatter to noise.

OTDR Ghosts

If you are testing relatively short cables with highly reflective connectors, you will likely encounter "ghosts." Ghosts are very confusing, since they seem to be real reflective events like connectors. Ghosts are simply multiple reflections from a highly reflective event back and forth between connectors. On very short cables, multiple reflections can cause multiple ghosts that are really confusing.

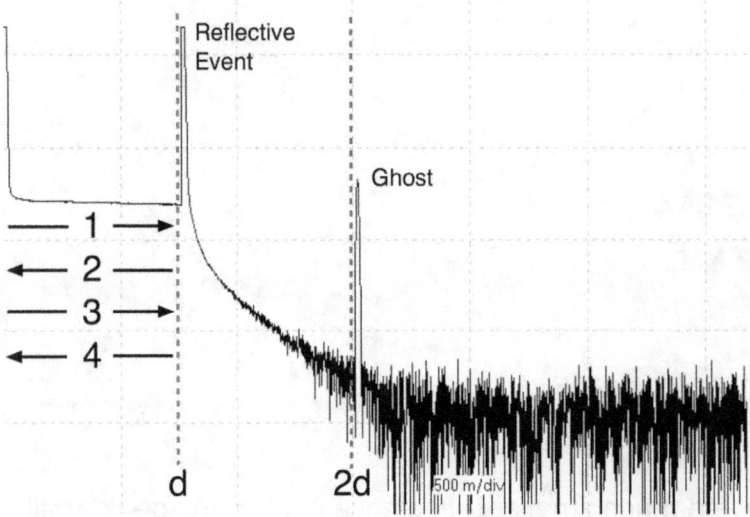

OTDR ghost from reflective end of a launch cable

Ghosts are caused by the strong reflectance from an event, typically a connector, reflecting back and forth in the fiber until it is attenuated to the noise level. In the drawing above, the four arrows show the path of the light causing the ghost. The arrow #1 shows the OTDR test signal going to the end of the cable where there is a connector. The connector is highly reflective and overloads the OTDR as shown by the flat top of the reflective peak.

The reflected light goes back to the OTDR as arrow #2 and shows on the trace as the event at distance "d." Some of the light going back to the OTDR is reflected from the connection on the OTDR back down the fiber, shown as arrow #3. That light is reflected again by the connector at distance d and goes back to the OTDR again, shown as arrow #4. That reflected peak shows up at distance 2d because the light has made two trips down the fiber.
We have seen traces from a cable that was tested with an
OTDR and deemed bad because the trace looked like the one above and the operator thought it was broken in the middle. The tester had not looked at the distance scale or he would have noted the "break" was at the length of the cable and the ghost at was exactly twice that length.

There are several ways to tell that a reflective peak is a ghost. First it will be in a locations where there is not supposed to be a connection. Secondly, the ghost will be at a distance that is a multiple of the distance to a high reflectance event. Thirdly, the ghost will not show any loss.

Stress Loss Or Splice Loss?

OTDRs are valuable troubleshooting tools and can find many problems in an installed cable plant. Stress in a cable caused by tight bends or kinks can cause both short term issues due to the extra loss and long term issues with reliability. But stress losses can be confusing on an OTDR trace if you do not have good documentation on the cable plant because it can look just like a fusion splice, a loss where there is no reflectance. However there is a way to distinguish the two.

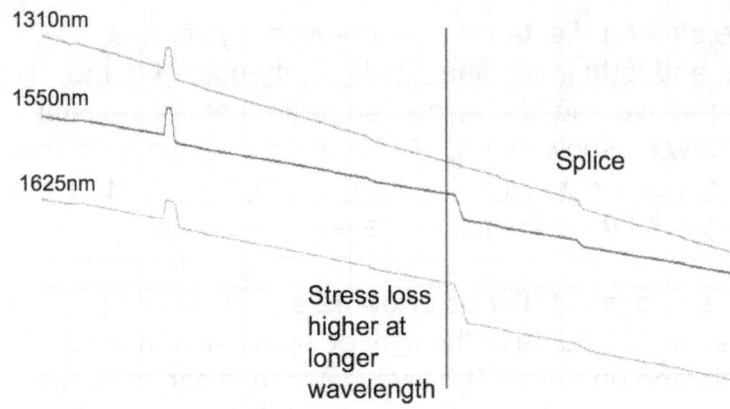

Traces of stress loss and splice events

The easy way to tell the difference in the two events is by making measurements at several wavelengths. Singlemode OTDRs should have test sources at two wavelengths, 1310 and 1550 nm, and many new units also have a source at 1625 nm. Stress on a cable will cause fiber losses that are higher at longer wavelengths, so taking traces at several wavelengths will make it easy to distinguish the stress loss which has higher loss at longer wavelengths from the splice which has similar losses at all wavelengths.

Comparing traces from two or three wavelengths will visually show where any stress problems exist in the cable plant. Note that sometimes these stress events occur in splice closures, so the traces will indicate that the closure needs opening and checking for fiber routing inside the closure.

Note: More information on OTDR measurement accuracy will be found in Chapter 14.

Documentation

Like every step of the fiber optic design, installation and operation processes, it is important to fully document the test and record all relevant data. All tests should be recorded with the following data as a minimum:

For all tests performed at one time:
- Date of the test
- Location of the test
- Environmental conditions (temp/humidity/local conditions)
- Cable plant identification (cable type/fiber type/connector type/ length)

- Type of test (visual inspection, insertion loss, OTDR, CD, PMD, SA)
- Test equipment used (type, brand, model, serial number, date of last calibration)
- Wavelength
- Insertion loss: reference method (1/2/3 cable reference methods)
- OTDR: manual or auto test
- Criteria for pass/fail (loss budget criteria)

For each individual test
- Identification of component under test (e.g. fiber #)
- Test results (actual data and pass/fail if noted)
- Note if results are filed electronically

Chapter Exercises

If you have an OTDR, use it for these exercises. If not, go to the FOA website and download the FOA OTDR simulator. Before you begin any exercises, read the OTDR manual or the OTDR simulator manual
.

Using a long length of cable with several splices or connections, make measurements in the OTDR auto test mode and record the setup parameters used for auto test. Save the trace. Then using manual setup, vary the test parameters and see if the results are better/worse/just different.

Measure a connector loss with the 2 point method and see how the loss varies with marker placement. Choose what you think is the best reading. Then repeat the test using LSA and compare the data.

Test a cable plant with an OTDR with both launch and receive cables and then test the cable with a test source and power meter or OLTS and compare the measurements.

Do a bidirectional test on two slightly different cables that show a gain in one direction and a larger loss in the other and use the method above to calculate the actual loss. When making the measurements, also measure the attenuation of the two fibers and analyze according to the description above.

Chapter Quiz

1. Length measurements by the OTDR are shorter than the actual cable because the fiber is longer than the cable itself.
    True
    False

2. Reflections seen in traces from OTDRs comes from mismatches in the index of refraction at a joint being tested.
    True
    False

3. Standards allow OTDRs to be used instead of a test source and power meter or OLTS (optical loss test set) to test a cable plant.
    True
    False

4. OTDR measurements are directional and traces should be taken in both directions and averaged to get reliable data.
    True
    False

4. "Ghosts" are caused by high reflectance events, usually connectors, in a short fiber optic cable.
    True
    False

6. OTDRs utilize the _____ in the fiber to make measurements.
    A. Bandwidth
    B. Backscatter
    C. Absorption
    D. Dispersion

7. OTDRs measure the length of a cable. What information is needed to make this measurement accurately?
    A. Attenuation of the fiber
    B. Speed of the light in the cable
    C. Whether the fiber is multimode or singlemode
    D. Connection and splice losses

8. Length measurements by the OTDR are about _____ than the actual cable because the fiber is loosely wound in the cable for protection from tension during pulling.
   A. 1-2% longer
   B. 2-5% longer
   C. 1-2% shorter
   D. 2-5% shorter

9. To make measurements of the entire cable being tested, OTDRs must be used with _____.
   A. A launch cable to measure the connector on the near end of the cable
   B. A receive cable to measure the connector on the far end of the cable
   C. Both launch and receive cables to include connectors on both ends of the cable

10. For testing a cable, the OTDR range should be set _____.
   A. As short as possible
   B. Approximately twice the expected length of the cable
   C. As long as possible
   D. Any of the above

The following questions refer to this diagram of an OTDR trace.

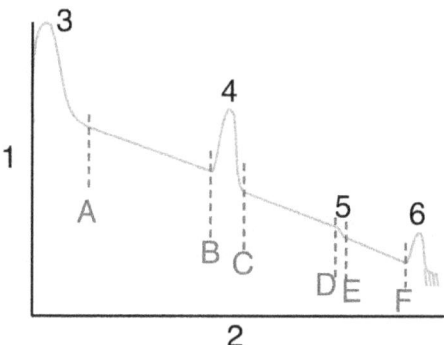

11. The OTDR trace shows a graph of the data in the format shown. The vertical (Y) axis shows _____.
   A. Optical power in dB
   B. Optical power in dBm
   C. Distance in meters, miles or kilofeet
   D. Time to the events

12. The horizontal (X) axis on an OTDR trace shows _____.
   A. Optical power in dB
   B. Optical power in dBm
   C. Distance in meters, miles or kilofeet
   D. Time to the events

13. The distance from the Y axis to point A on the trace that includes event 3 shows the _____.
   A. Length of the launch cable
   B. Power of the OTDR test pulse
   C. The dead zone of the OTDR
   D. Loss in the launch cable

14. From the Y-axis to point B is the length of the _____.
   A. Dead zone
   B. Launch cable
   C. Cable under test
   D. Test pulse width

15. To measure the length of the cable under test, you need to measure the distance between points _____ and _____.
   A. A and B
   B. C and F
   C. A and F
   D. B and F

16. Event #4 is a connector since it shows _____.
   A. Splice
   B. Both reflectance and loss
   C. Kink in the cable
   D. Break in the cable

17. Event #5 shows loss but no reflectance, so it is either a _____ or a _____.
   A. Fusion splice
   B. Prepolished/splice connector
   C. Kink in the cable
   D. Break in the cable

18. Measuring from point B to point C on the trace will give the _____ at event #4.
   A. The length of the connector
   B. Loss of the connector
   C. Reflectance of the connector
   D. Dirt on the connector

19. To measure the attenuation coefficient of a segment of the cable in dB/km, you should measure between points _____ and ____.
    A. A and B
    B. C and D
    C. E and F
    D. All of the above

20. Event #6 is the connector on the end of the cable. How can you measure the loss of this connector?
    A. Place a marker after the pulse and measure the loss
    B. Measure the height of the reflectance pulse
    C. Measure the width of the reflectance pulse
    D. Add a receive cable connected to the end of the cable under test

Chapter 10

Fiber Characterization

Objectives: From this chapter you should learn:
What Fiber Characterization involves
What are Chromatic Dispersion (CD) and Polarization Mode Dispersion (PMD)
How CD and PMD affect high speed long distance transmission
What is Spectral Attenuation (SA)
How SA affects wavelength-division multiplexing
How one tests CD, PMD and SA

Fiber Characterization

One of the big advantages of fiber optics is its capability for long distance high-speed communications. Singlemode fiber attenuation at long wavelengths (~1550 nm) is extremely low. Fibers can be fusion spliced with virtually no loss. High-powered lasers, sophisticated transmission protocols and fiber amplifier regenerators mean long distances are easily obtained. Dense wavelength division multiplexing (DWDM) allows up to 128 channels of signals on a single fiber.

However, for high-speed networks operating over very long distances, new factors limiting fiber performance become important. Chromatic dispersion, the dispersion caused by light of different wavelengths, and polarization mode dispersion, caused by the polarization of the light in the fiber, become factors limiting the bandwidth capacity of fiber links. Pulse broadening due to chromatic dispersion and the variation of fiber attenuation with wavelength can become issues with DWDM.

All these factors need testing on long distance networks to ensure proper link performance. Tests are performed on new installations to ensure the fiber being installed is capable of future upgrades. Older cable plants are tested to evaluate fibers for upgrades of legacy communications systems at slower speeds.

A suite of tests for these factors has been developed to test fibers for long distance high-speed networks. These tests are normally called "fiber characterization," but technically they are "fiber optic cable plant characterization" since it must include the complete end-to-end cable plant.

These are *NOT* the only important tests for long links, they are in addition to the traditional cable plant tests: careful inspection of connectors and the installed cable plant (neatness and lack of stress in cables and patchcords), insertion loss testing with a test source and power meter or optical loss test set (OLTS) and optical time domain reflectometer (OTDR) testing. You can review those tests in the earlier chapters. In this chapter we will describe the additional tests.

Tests For Fiber Characterization

Test	Reason For Test
Connector Inspection	Verify quality and condition of connectors
Insertion Loss	Compare cable plant loss to loss budget and network power budget
ORL/Reflectance	Look for locations of reflectance problems
OTDR	Check splices, connectors, fiber attenuation, look for stress induced losses from installation
Spectral Attenuation	Wavelength division multiplexing uses fiber over a large range of wavelengths
Chromatic Dispersion	Long distances at high speeds (>2.5Gb/s) may suffer dispersion
Polarization Mode Dispersion	Long distances at high speeds (>2.5Gb/s) may suffer dispersion

Background Information

Here is some background information that is commonly used in defining OSP links and testing. This can be used as reference material for understanding standards, tests and reports.

Fiber Types
These are the fiber types with cross references to the nomenclature for TIA, IEC and ITU that can be confusing. In addition, there are subcategories of most of the ITU G.65x types for specialized applications. The ITU has defined a series of specifications that describe the geometrical properties and transmission properties of multimode and single-mode fiber-optic cables that are widely used for testing.

These are the types of SM fibers available by international designations

Fiber Type (TIA/IEC/ITU)	Description	Applications
OS1/B1.1/G.652	Standard SM fiber for 1310nm	Original SM fiber - optimized for 1310 nm - OK for use at 1550
OS2/B1.2/G.652	Low water peak fiber	Processed to reduce water peak absorption for DWDM
---/B2/G.653	Dispersion-shifted fiber	Optimized for 1550 nm
---/B1.2/G.654	Cutoff-shifted fiber	Optimized for low loss at 1500 to 1600 nm for long haul, submarine cables
---/B4/G.654	Non-zero dispersion-shifted fiber	Optimized for low loss at 1500 to 1600 nm for long haul, submarine cables
---/---/G.655	Non-zero dispersion-shifted fiber	Optimized for 1550 nm - DWDM
---/---/G.656	Wideband Non-Zero Dispersion-Shifted Fiber	Wideband - DWDM from 1460 to 1625 nm
---/---/G.657	Bend-insensitive fiber	Includes all types of fibers listed above

Wavelength Bands
It is common to see references to wavelength bands for fiber optic systems. These can be confusing unless you know the relevant wavelengths and uses.

Wavelength bands used in fiber optic communications

Wavelength Band	Wavelength Range (nm)	Definition
O-band	1260-1360	Original band, PON upstream
E-band	1360-1460	Water peak band
S-band	1460-1530	PON downstream
C-band	1530-1565	Lowest attenuation, original DWDM band, compatible with EDFA fiber amplifiers, AM CATV
L-band	1565-1625	Low attenuation, expanded DWDM band
U-band	1625-1675	Ultra long wavelengths

Dense wavelength division multiplexing (DWDM) originally used optical signals multiplexed within the 1550 nm band compatible with erbium doped fiber amplifiers (EDFAs), which are effective for wavelengths between approximately 1525–1565 nm (C band), or 1570–1610 nm (L band). Dense wavelength division multiplexing (DWDM) channel plans vary, but a typical system might use 40 channels at 100 GHz spacing or 80 channels with 50 GHz spacing.

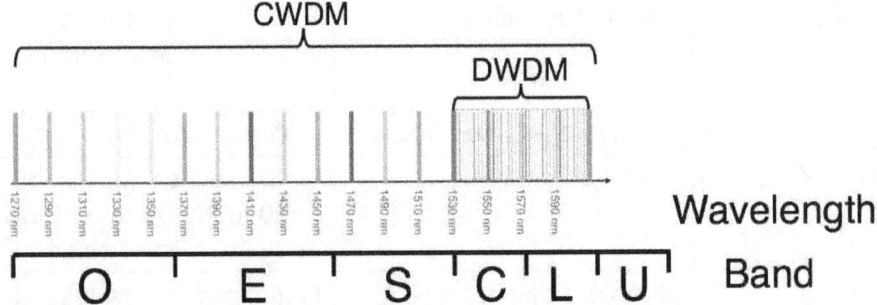

Wavelength division multiplexing in the transmission bands

Coarse wavelength division multiplexing (CWDM), a lower cost version of WDM, uses lasers from 1260 to 1670 nm in 20 nm windows. This allows less expensive lasers but does require verification of the spectral attenuation of the fiber over a longer range of wavelengths.

Spectral Attenuation (SA) - Also Called Attenuation Profiling (AP)

With the development of low water peak fibers, the possibility of transmission from 1260 to 1675 nm has been realized. Since one may want to use available fibers of unknown spectral attenuation for DWDM or CWDM that use spectrum all the way from 1260 to 1670 nm, it becomes necessary to test for spectral attenuation to verify the usability. At the water peaks (1244 and 1383 nm – there is also a peak out of band at 950 nm), legacy fibers may have attenuation coefficients around 2 dB/km while low water peak fibers may be as low as 0.4 dB/km.

Typical Fiber Attenuation Over Wavelength Range (Corning SMF 28)

Wavelength (nm)	Maximum Loss (dB/km)
1310	0.33 – 0.35
1383	0.31 – 0.35
1490	0.21 – 0.24
1550	0.19 – 0.20
1625	0.20 – 0.23

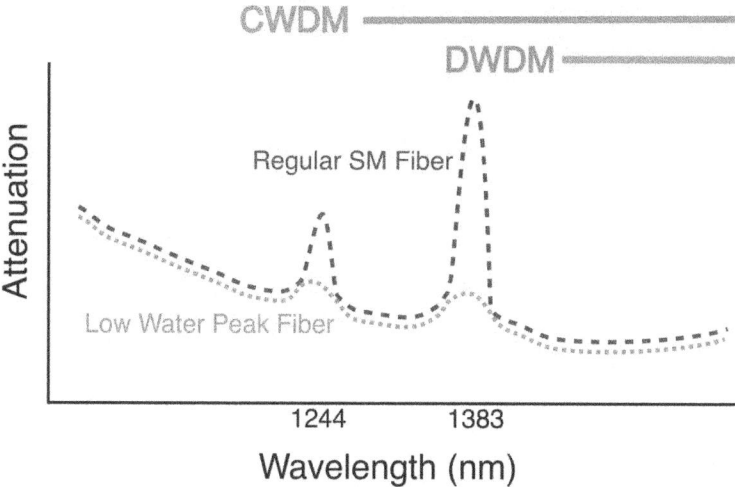

Spectral attenuation of regular and low water peak singlemode fibers

Spectral attenuation should be tested in the wavelength ranges (bands) of interest. DWDM systems focus mainly on the C-band, with some also operating the S- and L-bands. CWDM systems operate in the entire singlemode wavelength range, 1260 to 1625 nm.

Testing spectral attenuation is done per TIA/EIA-455-61 or IEC 61300-3-7 with broadband sources like LEDs and a spectrum analyzer on the receiving end of the fiber. Calibration is done with a short fiber length and then the instrument calculates the spectral attenuation on a long length being tested. The measurement of spectral attenuation uses instruments similar to those used for CD testing so some instruments do both measurements at one time.

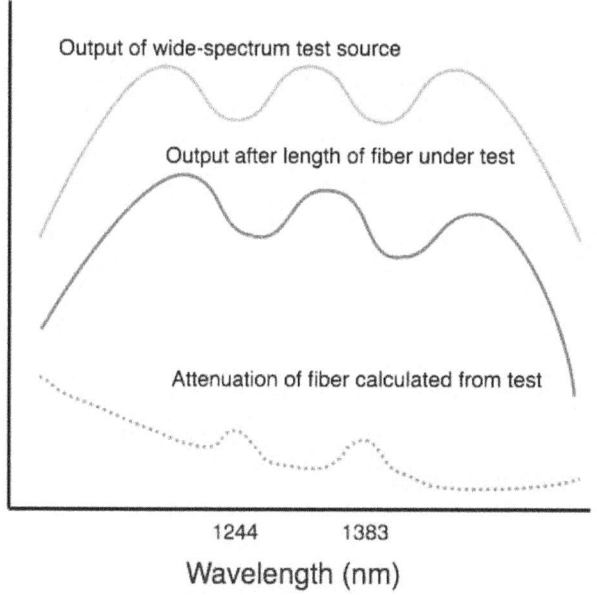

Measuring spectral attenuation with wideband spectral source

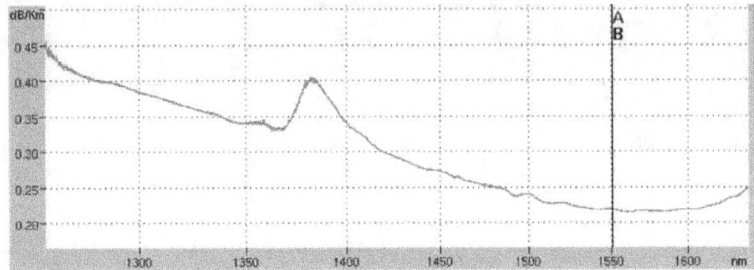

Actual data from spectral attenuation test

Chromatic Dispersion (CD)

Chromatic dispersion (CD) is caused by the fact that singlemode glass fibers transmit light of different wavelengths at different speeds. The ratio of the speed of light in a medium to the speed in a vacuum defines the index of refraction or refractive index of the material. For optical fiber, the effective index of refraction is about 1.45, so the speed of light in glass is about 2/3 the speed of light in a vacuum. But the index of refraction, and thereby the speed of light in the fiber, is a function of the wavelength of light, the principle we all know from seeing a prism break light into a spectrum.

Most sources used in long distance fiber optic links are lasers that have very little spectral width and fibers are optimized for the wavelength of use. Both these factors minimize the effects of chromatic dispersion but cannot totally stop it. Singlemode fiber optic links use lasers for transmitters. Slower systems use Fabry-Perot (F-P) lasers that have a relatively narrow spectral width (range of wavelengths in the source). Faster systems and DWDM systems use distributed feedback (DFB) lasers that have spectral widths less than 1/10th as large as F-P lasers.

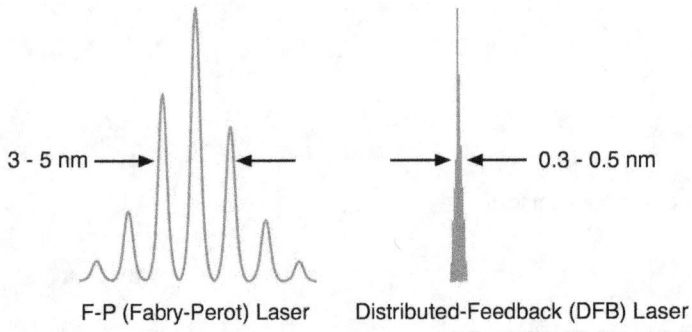

F-P (Fabry-Perot) Laser   Distributed-Feedback (DFB) Laser

Relative spectral width of F-P and DFB lasers

As the pulse proceeds down the fiber, the light of longer wavelength travels slightly faster and spreads the pulse out as shown here.

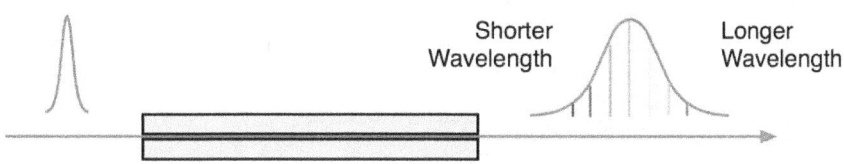

Chromatic dispersion in fiber

Causes Of Chromatic Dispersion
There are two factors that cause chromatic dispersion in the fiber: material dispersion and waveguide dispersion.

Material Dispersion
Material dispersion is caused by the variation of the index of refraction in a given material, glass in this case, over wavelength. Looking at the graph below, the variation of the index of refraction over the entire spectrum covered by fiber optics may seem small, only a few percent, but when you are dealing with very high speed pulses over very long distances it can add up.

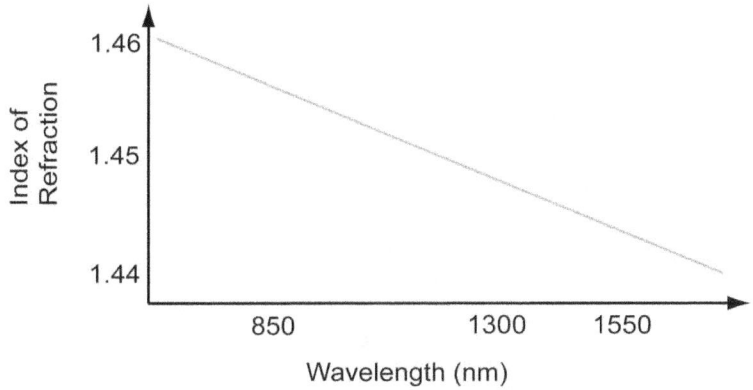

Material dispersion in optical fibers

Waveguide Dispersion
Waveguide dispersion is a bit more complex. In singlemode fiber, the wavelength of the light is not that much bigger than the core of the fiber and as a result the light traveling down the fiber actually travels in an area that exceeds the diameter of the core, which is called the "mode field diameter" of the fiber. The mode field diameter is a function of the wavelength of the light, with longer wavelengths traveling in a larger mode field diameter.

Thus part of the light in a signal is traveling in the geometric core of the fiber and part is traveling in the cladding. Since the core is made of a higher index of refraction glass than the cladding, the light in the cladding travels faster than the light in the core. Longer wavelengths have larger mode field diameters so they suffer more waveguide dispersion. This variation due to wavelength causes chromatic dispersion.

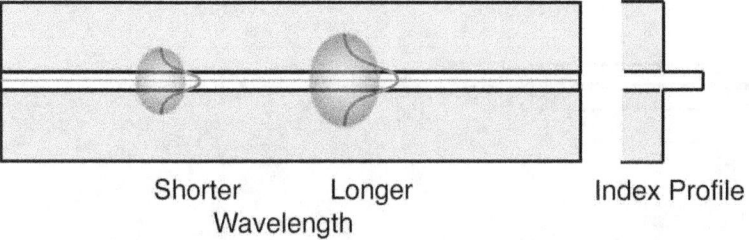

Shorter   Longer   Index Profile
       Wavelength

Waveguide dispersion in singlemode optical fiber

Engineered Dispersion in Fibers
Material and waveguide dispersion have different variations with wavelength, so careful design of the fiber materials and index profiles allows the fiber to have a "zero dispersion wavelength." On either side of that wavelength, dispersion increases, positive at higher wavelengths and negative at lower wavelengths.

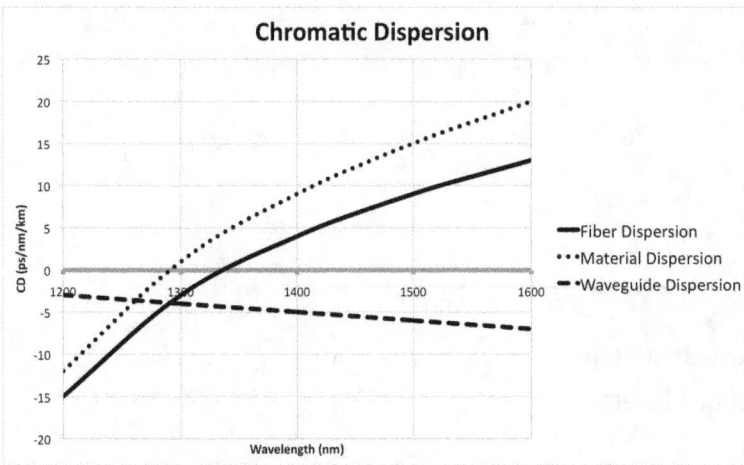

Elements contributing to fiber chromatic dispersion

The importance of chromatic dispersion is a function of the application for the fiber. As a result, different SM fibers have been developed for the requirements of specific applications.

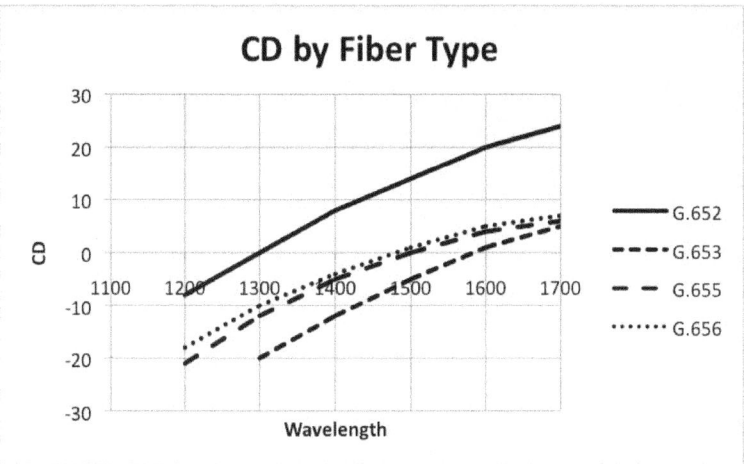

Chromatic dispersion in different singlemode fiber types

For each fiber, the specifications for CD are the absolute dispersion and for some fibers like G.652, the slope of the dispersion curve at the zero dispersion wavelength (λ_o). G.652 fiber is widely used for short and medium length networks. This fiber can be used for transmission at both 1310 nm and 1550 nm, so it is specified for CD at both wavelengths. In fact, the zero dispersion wavelength is around 1310 nm so chromatic dispersion is low, but at 1550 nm, the CD is high, but acceptable for many applications.

The units for specifying chromatic dispersion are the actual dispersion, in picoseconds (ps) per nanometer of spectral width (nm) per kilometer of fiber, or

CD = ps/nm-km)

In addition, fibers that may be used over a broad range of wavelengths like G.652 will have a specification for the slope of the CD curve over wavelength at the zero dispersion point ($\lambda_{0)}$. This slope is the derivative of the curve at λ_0, and the specification is expressed in picoseconds per nm² per kilometer:

$Slope_{\lambda 0}$ = ps/nm²-km

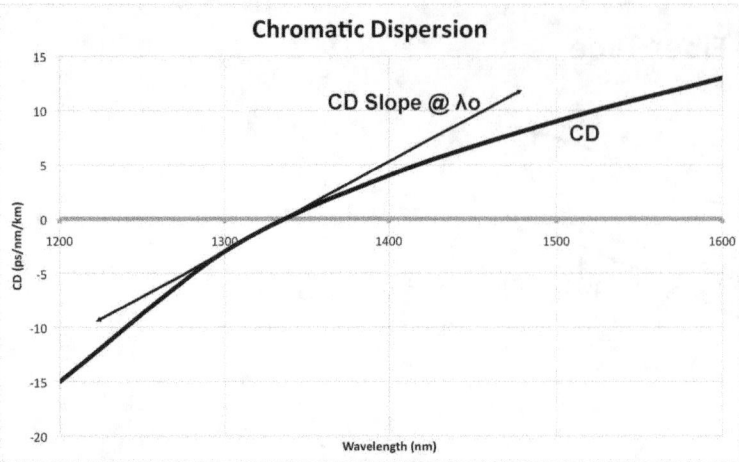

Chromatic dispersion fiber specification (G.652)

CD specifications for typical fibers

Fiber	λ_0 (nm)	Slope$_{\lambda 0}$ (ps/nm^2-km)	D$_\lambda$ (ps/nm-km)
G.652	1300-1324	0.092	17 @ 1550
G.653.A	1500-1600	0.085	3.5
G.655.A-C	1530-1565	NS	>1, <10*
G.655.D	1530-1565	NS	>1.2, <7.2*
G.655.E	1530-1565	NS	>4.8, <10.1*
G.656	1460-1550 1550-1625	NS	>2, <14*

*Approximate, calculated function of wavelength (λ), must be >1 but <10 or as noted

CD Budget
Here is an example of how to calculate an estimate of the CD of a cable plant – call it a "CD Budget."
Simply multiple the chromatic dispersion D_λ of the fiber by the length just like calculating a loss budget. This will give you an idea of what to expect and compare to test results, as well as compare to the requirements of the potential usage for higher speed networks shown below.

Calculate based on "D_λ" and fiber length, e.g.

50km G.652 fiber = 50km X 17ps/nm-km = 850ps/nm

100km G.653.A fiber = 100km X 3.5ps/nm-km = 350ps/nm

If testing is being done for a system upgrade, there may be a CD specification for the intended system type and speed. Below are some maximum CD specifications for some typical networks.

Typical System Specifications For Maximum Chromatic Dispersion

Bit Rate (Gb/s)	System	Max CD (ps/ nm)@1550nm	Maximum (km) G.652	Distance (km)G.655
2.5	SDH STM-16 SONET OC-48	18817	1100	4700
10	SDH STM-64 SONET OC-192	1176	70	290
10	Ethernet	738	44	180
40	SDH STM-256 SONET OC-768	73.5	5	20

Note the table stops at 40Gb/s. There are 100Gb/s systems and even faster being used or in field trials today. Some of those are WDM, e.g. 100 Gb/s with 10X10 Gb/s or 4X25Gb/s and some long distance ones that use coherent transmission. The best way to determine system limits is to ask the manufacturer of any specific system of interest.

Telcordia/Bellcore standard GR-253-CORE provides another method of estimating CD effects. The standard specifies that for a 1 dB performance penalty, the total dispersion should be less than 0.306 times one bit period. This can be expressed as:

D_λ (fiber dispersion) X L(length) X B (bit rate) X $\Delta\lambda$(source spectral width) < 0.306

Thus for a G.652 fiber with a D_λ =17 ps/nm-km, B=10Gb/s and $\Delta\lambda$=0.1nm, the maximum length would be:

(17×10^{-12}) L $(10 \times 10^{+9})$ (0.1) <0.306
17×10^{-3} L <0.306
.017 L <0.306
L < 18 km

Likewise you could insert a given length (for example on a cable plant being tested) and projected bit rate and calculate a maximum D_λ for a given fiber. Or you can solve for a given bitrate and calculate a maximum CD: D_λ x L.

Chromatic Dispersion Compensation

As mentioned earlier, the dispersion characteristics of a fiber can be manipulated by the materials and design of the fiber. In fact, fibers can be made that have CD inverse to the typical fibers and of a greater magnitude. So a length of dispersion compensation fiber can be added to a link, usually at a repeater (optical amplifier) that reverses the CD of the fiber span before it. Such fibers tend to have high loss and bend sensitivity, so alternatively a dispersion compensator made from a specialized component called a Bragg grating can be used, but it has a more limited use and higher cost.

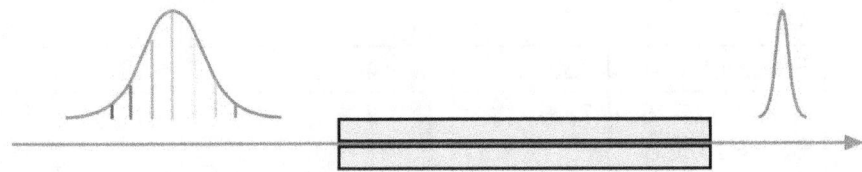

Dispersion compensation removes the effects of CD

Chromatic Dispersion in the Cable Plant

As with any other component, optical fiber performance parameters can vary from batch to batch, so a long concatenated cable plant with many different fibers will have an end-to-end chromatic dispersion which is an integration of the CD of all the individual fibers. Therefore fibers in long distance links will probably be tested for CD after installation or before upgrading a link to higher bit rate electronics.

Testing Chromatic Dispersion

There are several methods used for testing CD in fibers. All involve testing at a variety of wavelengths. The test sources can be several discrete sources of various wavelengths, a tunable laser or a broadband source with a monochromator in the receiver. The results of the tests are the relative speeds of the signals at the various wavelengths. The data taken at discrete wavelengths is then analyzed to calculate the dispersion in terms of ps/nm/km.

All these methods have international standards for the test methods, instruments and data analysis.

Standards	Description
IEC 60793-1-42	Measurement methods and test procedures—chromatic dispersion
ITU-T G.650.1	Definitions and test methods for linear, deterministic attributes of singlemode fiber and cable
TIA FOTP-175-B	Chromatic dispersion measurement of single-mode optical fiber
GR-761-CORE	Generic criteria for chromatic dispersion test sets

Test methods use phase delay or time of flight and generally require access to both ends of the fiber as well as a second fiber for synchronization of the two test instruments at either end. However, an OTDR test method is also used where traces are taken at several discrete wavelengths and CD can be calculated from the data obtained from the traces, allowing testing in the field from one end of the fiber.

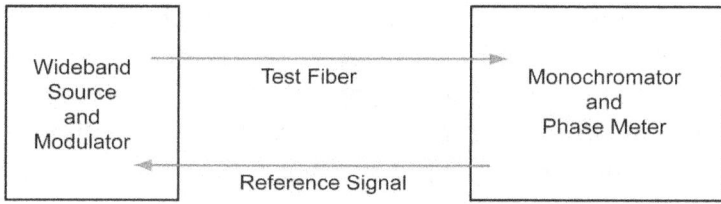

Phase shift method

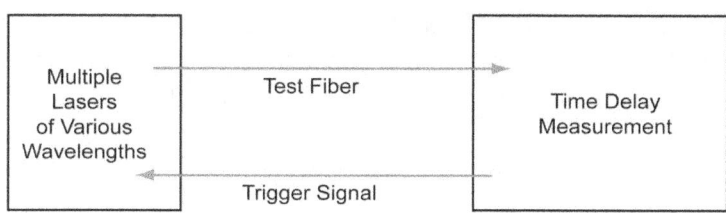

Pulse delay method

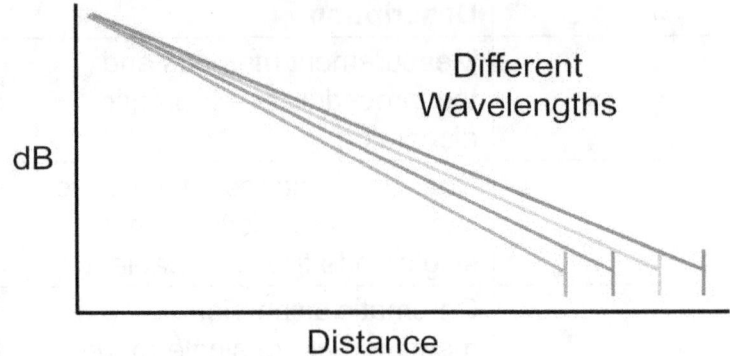

Distance

OTDR test method requiring access to only one end of the fiber

Test results will provide a total CD for the link and perhaps D_λ and the slope at λ_o.
As with all complex tests, it is important for the test tech to understand the tests and expected results plus how to operate the test equipment properly and analyze the data.

Polarization Mode Dispersion (PMD)

Polarization mode dispersion (PMD) is a bit more complex than chromatic dispersion. Polarization is a phenomenon of light traveling in a medium as an electromagnetic wave with components at right angles. Some materials, like a glass optical fiber, have a different index of refraction for each of those components of the light wave, which is called birefringence.

A different index of refraction means light travels at a different speed, so in the simplest visualization, PMD in fiber looks like the drawing below where each component of the polarized light travels at a different speed, causing dispersion. The magnitude of PMD in a fiber is expressed as this difference, which is known as the differential group delay (DGD) and called Δτ ("delta Tau").

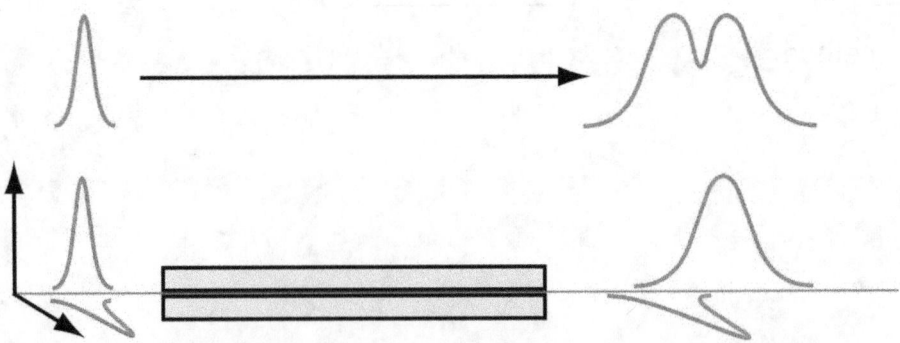

Polarization mode dispersion in singlemode fiber

Causes of Polarization Mode Dispersion

PMD is caused by the birefringence of the fiber that can be influenced by two factors, material birefringence and waveguide birefringence. Waveguide birefringence is caused by geometrical variations in the fiber such as concentricity or ellipticity. Material birefringence is mainly caused by stress on the fiber.

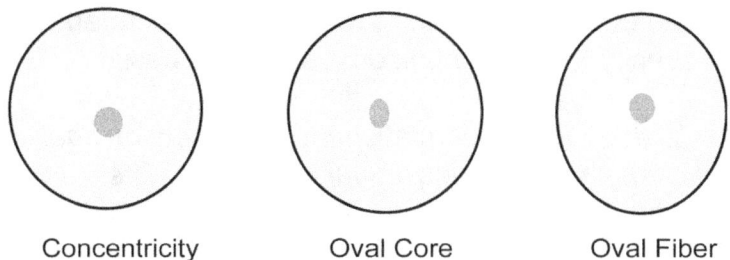

| Concentricity | Oval Core | Oval Fiber |

Waveguide Birefringence

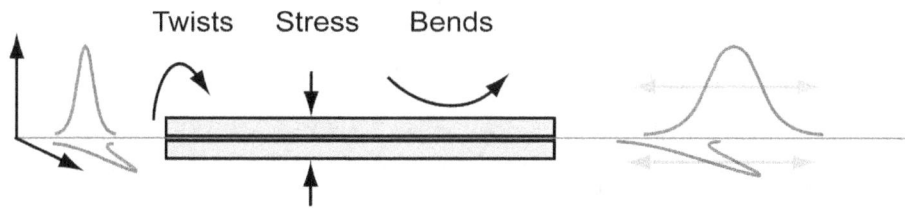

Material birefringence

What is measured for PMD is basically the variation in time of the two polarization modes and the dispersion that variation creates. PMD is tested for the entire cable plant over a range of wavelengths.

PMD is a complex issue in installed optical fiber. In a long concatenated fiber, each fiber in the link can have different waveguide and material birefringence characteristics. This is caused by the random waveguide characteristics of each fiber in the link and variations of the stress on each fiber along the length of the cable. The measured PMD will be a complex mix of the effects on all fibers connected in the cable plant.

PMD also varies over time as the environment of the cable plant changes – temperature, vibration, stress by tension or bending, etc. can affect the cable plant's PMD. Variations are particularly noticeable in aerial fiber, where the PMD may vary considerably according to temperature and wind speed buffeting the fiber. Underground fiber near sources of vibration like railroad tracks may show variations in PMD also.

Unlike CD that is a constant and relatively easy specification to test, PMD needs to be tested over a period of time and the resulting data analyzed for average and peak values.

Fiber Specifications For PMD

Unlike most fiber specifications, PMD is not a concrete specification, but is tested and specified on a statistical basis for cabled fiber. Cabled fiber, of course, has the stress induced due to the cabling process. Manufacturers of transmission systems also specify PMD for a link on a statistical basis.

From ITU G.652 Standard: *"Cabled fibre polarization mode dispersion shall be specified on a statistical basis, not on an individual fibre basis. The requirements pertain only to the aspect of the link calculated from cable information."*

"The manufacturer (of transmission equipment) shall supply a PMD link design value, PMDQ, that serves as a statistical upper bound for the PMD coefficient of the concatenated optical fibre cables within a defined possible link of M cable sections. The upper bound is defined in terms of a small probability level, Q, which is the probability that a concatenated PMD coefficient value exceeds PMDQ. "

Typical Fiber Specifications For PMD

Fiber Type	PMD$_Q$
G.652A/C	0.50 ps/√km
G.652B/D	0.20 ps/√km
G.653A	0.50 ps/√km
G.653B	0.20 ps/√km
G.655C/D/E	0.20 ps/√km
G.656	0.20 ps/√km
G.657A	0.20 ps/√km
G.657B	0.50 ps/√km

From TIA FOTP-122 (an adaptation of IEC 61282-9):

"In long fibre spans, DGD (differential group delay caused by birefringence) is random in both time and wavelength since it depends on the details of the birefringence along the entire fibre length. It is also sensitive to time-dependent temperature and mechanical perturbations on the fibre. For this reason, a useful way to characterize PMD in long fibres is in terms of the expected value, <Δτ> (ps/√km), or the mean DGD over wavelength. In principle, the expected value <Δτ> does not undergo large changes for a

given fibre from day to day or from source to source, unlike the parameters δτ or Δτ. In addition, <Δτ> is a useful predictor of lightwave system performance.

Table from the ITU G.652/G.653 fiber specs for PMD limits

Max PMD	Link Length	Max Fiber DGD	Bit Rate
Not Specified			Up to 2.5Gb/s
0.5 ps/√km	400	25 ps	10Gb/s
	40	19 ps	10Gb/s*
	2	7.5 ps	40Gb/s
0.20 ps/√km	3000	19 ps	10Gb/s
	80	7.0	40Gb/s
0.10 ps/√km	>4000	12 ps	10Gb/s
	400	5 ps	40Gb/s

* Also applies to 10Gb/s Ethernet systems

The table below shows examples of the distance limitations for some typical systems. Here is how it is calculated.

The bit period is calculated from the bit rate – it is the period from bit to bit, so 1Gb//s means 1 bit per nanosecond, 10 Gb/s means 1 bit per 100nanoseconds, etc.

The Max Mean DGD (Δτ) is calculated by a simple estimate – it should be no larger than 1/10 of the bit rate to not cause problems.

The DGD varies by the square root of the length so, for 2.5Gb/s, the bit period is 400ps, the max DGD is 40ps (1/10 of 400ns), and for 400km, the √400=20 so we calculate 40ps/20=2ps/√km.

Note the sensitivity of networks to PMD rises linearly with bit rate and with the square root of length. Ethernet has a lower tolerance to PMD than SDH/SONET because it has a different error correction scheme and a BER requirement that is more stringent than SDH/SONET. The numbers in the table below are estimated as the requirements are still under study.

Bit Rate (Gb/s)	System	Bit Period (ps)	Max Mean DGD ($\Delta\tau_{max}$) (ps)	PMD Coeff (ps√km) @ 400km
2.5	SDH STM-16 SONET OC-48	400	40	<2
10	SDH STM-64 SONET OC-192	100	10	<0.5

| 10 | 10G Ethernet* | 100 | 5 | --- |
| 40 | SDH STM-256 SONET OC-768 | 25 | 2.5 | <0.125 |

Here is a graphical way of looking at the data on bitrate, PMD and distance. This helps visualize the approximate PMD coefficient for each network at various distances. Engineers designing these systems will look at each network separately, including other factors like the data encoding formats and fiber specs. Consider this and the previous slide approximations only.

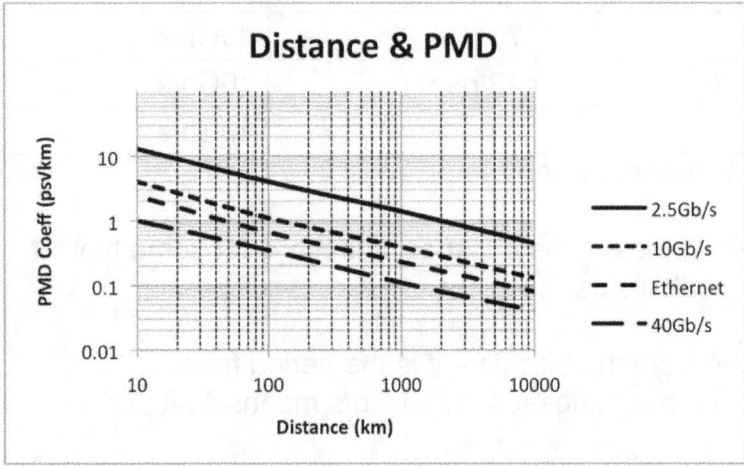

PMD effects on distance at different bitrates

PMD causes pulse broadening and/or jitter in the received electrical signal, potentially causing errors in the reception of the signals. Since the PMD can vary over time, an extra margin of 1 to 3 dB is often added to the power budget to accommodate variation in PMD.

PMD is an important issue as data rates on long distance links increases to 40 Gb/s, 100 Gb/s and above. Unfortunately, there appear to be no reliable compensation schemes for PMD, so the only solution is to test links to be upgraded for PMD using one or more of the standardized test methods.

Expected Test Results (PMD Budget)
Here is a table based on the earlier table from the ITU G.652/G.653 fiber specs for PMD limits. Instead we show how to calculate the DGD from the average PMD specification for a given fiber length. This provides a reference value for each link when testing in the field.

PMD (ps/√km) X √length (km) = DGD (ps)

Calculating DGD From Average PMD

Max PMD	Link Length (km)	√km	DGD (ps)
0.5 ps/√km	400	20	10ps
	40	6.3	3.2ps
	2	1.4	0.7ps
0.20 ps/√km	3000	54.8	11ps
	80	8.9	1.8ps
0.10 ps/√km	4000	63.2	6.3ps
	400	20	2ps

Testing PMD

PMD is generally tested for fibers during manufacture and after being cabled. In the field, it is common to test PMD on newly installed fibers which are intended for operation at high speeds - generally above 2.5 Gb/s. PMD may also be tested when considering upgrading fibers installed some time in the past to faster networks. Since PMD varies over time, a single test becomes an average and tests at a later time may be done for comparison.

There are a number of commonly used test methods for PMD, some of which are limited to the manufacturing environment, while others can be used in the field. Essentially, all the test instruments have a source that can vary the polarization of the test signal and a measurement unit that can analyze polarization changes.

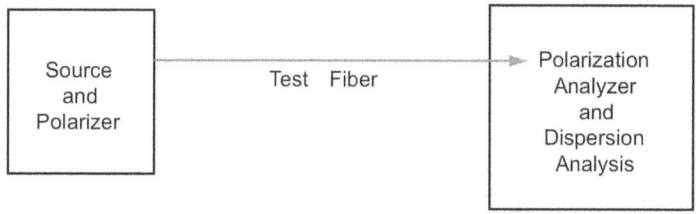

Basic PMD test configuration

Here are descriptions of the methods and relevant standards.

Description	Test Method	Standards
PMD for single-mode optical fibers by the Fixed Analyzer method	Extrema Counting (EC) Fourier Transform (FT)	FOTP-113

PMD measurement for single-mode optical fibers by Stokes parameter measurements	Jones Matrix Eigen-analysis (JME) Poincare Sphere Analysis (PSA)	FOTP-122
PMD measurement for single-mode optical fibers and cable assemblies by Interferometry	Traditional Interferometry(TINTY) General Interferometry (GINTY)	FOTP-124
Guideline for PMD and DGD measurement in single-mode fiber optic components and devices		FOTP-196
Measurement methods and test procedures—polarization mode dispersion	Fixed Analyzer measurement method (EC / FT) Stokes evaluation method (JME / PSA) Interferometry method (TINTY)	IEC 60793-1-48
Portable PMD test sets used for analyzing single-mode fiber		GR-2947-CORE
Definitions and test methods for statistical and nonlinear attributes of single-mode fiber and cable	Stokes parameter evaluation technique (JME & PSA) State of Polarization method (SOP) Interferometric methods (TINTY & GINTY) Fixed Analyzer technique (EC / FT / Cosine Fourier Analysis)	ITU-T G.650.2
OTDR-based single-ended test method (EXFO patent) Isolates PMD by fiber section along a concatenated cable. EXFO equipment tests both CD and PMD	Scrambled State-of-Polarization Analysis (SSA) Random scrambling polarization optical time domain reflectometry (RS-POTDR).	IEVC 61282-9 CD: EIA/TIA FOTP-175B; PMD: EIA/TIA FOTP-243.

Whatever method the test equipment uses, the data will probably be presented as DGD in ps with its wavelength dependence, perhaps as a graphical display like a spectrum.

PMD testing is not an easy, reproducible, accurate test. The measurement uncertainty can be as high as 10-20%, as shown by testing done within international standards committees. These committees have concluded that all these measurement techniques are permissible, that there are factors in making these measurements that are not well understood, and the methods of data analysis are not without question.

All this uncertainty of PMD measurements has the effect of making comparisons between tests and test methods difficult. Variations are particularly high on tests of older fiber links. Field measurements have even shown that variations in PMD can be correlated to environmental factors such as wind speed for aerial cables. PMD tests are often done over long time periods to average data and look for periods of high PMD that could affect transmission.

PMD testing needs to be done on all long high-speed links but the data must be analyzed intelligently to be of real use.

PMD Compensation
Compensation for PMD has been studied and discussed for years but no simple, reliable compensation method has been offered.

Fiber Characterization Test Equipment

There are a number of test sets available for lab and field testing of the parameters we combine in fiber characterization. For field testing, test sets may combine several of these tests into one instrument. Some require remote instruments for operation and some, primarily those based on OTDR test techniques, offer single-ended testing.

Due to the complexity of these tests and the instruments involved, as well as the evolution of the market to provide tests for increasingly higher bitrate systems, it is beyond the scope of this chapter to cover these instruments in the depth of the coverage of OLTS and OTDRs, for example.

As with all test equipment, it is important to first understand the tests involved and then to choose appropriate test equipment. Once equipment is chosen the operators should get trained on using that instrument so they know how to use it to perform the tests correctly and how to interpret the data it provides in the context of those tests.

Documentation

Like every step of the fiber optic design, installation and operation processes, it is important to fully document the test and record all relevant data. All tests should be recorded with the following data:

For all tests performed at one time:
- Date of the test
- Location of the test
- Environmental conditions (temp/humidity/local conditions)

- Cable plant identification (cable type/fiber type/connector type/ length)
- Type of test (visual inspection, insertion loss, OTDR, CD, PMD, SA)
- Test equipment used (type, brand, model, serial number, date of last calibration)
- Wavelength
- Insertion loss: reference method (1/2/3 cable reference methods)
- OTDR: manual or auto test

For each individual test
- Identification of component under test (e.g. fiber #)
- Test results
- Note if results are filed electronically

Chapter Exercises

Chromatic Dispersion
- Determine the allowable CD for a system
- Calculate the CD for several cable plants
- Test CD of several fiber samples
- Concatenate several links and see how the total tests compared to each segment
- Change concatenation – move fibers around and retest
- If you have dispersion compensating fiber, add it to the concatenated link and test again, compare results

Polarization Mode Dispersion
- Determine or calculate the allowable PMD for a system
- Test PMD of several sample cable plants
- Concatenate several links and see how the total tests compared to each segment
- Change concatenation – move fibers around and retest
- Stress the fiber in various locations and examine results

Chapter Quiz

1. Fiber characterization includes CD, PMD and SA tests in addition to the usual connector inspection, insertion loss and OTDR tests.
    True
    False

2. All fibers intended for use at speeds above 1Gb/s require fiber characterization.
> True
> False

3. Polarization mode dispersion PMD) can be affected by environmental factors like wind speed on aerial cable or vibration from railroads on buried cable.
> True
> False

4. PMD effects can be compensated by using special fibers near the receiver.
> True
> False

5. _____ needs testing on long distance fiber networks to ensure proper link performance at high speeds.
> A. Chromatic dispersion
> B. Polarization mode dispersion
> C. Insertion loss
> D. All of the above

6. The development of _____ allowed the use of CWDM (coarse wavelength division multiplexing) over the full range of singlemode fiber bands.
> A. Bend-insensitive fiber
> B. Multicore fiber
> C. Low water peak fiber
> D. Tunable lasers

7. The _____ test for CD requires access to only one end of the fiber.
> A. Phase shift
> B. Pulse delay
> C. Aggregate
> D. OTDR

8. Waveguide dispersion in singlemode optical fiber is caused by the difference in _____ of the fiber at different wavelengths.
> A. Mode field diameter
> B. Attenuation
> C. Backscatter
> D. Index of refraction

9. Material dispersion in singlemode optical fiber is caused by the difference in _____ of the fiber at different wavelengths.
   A. Mode field diameter
   B. Attenuation
   C. Backscatter
   D. Index of refraction

10. Stress on the fiber can cause variations in _____ in cabled singlemode fiber.
    A. Chromatic dispersion
    B. Polarization mode dispersion
    C. Mode field diameter
    D. Spectral attenuation

11. Fiber characterization test results should be compared to _____ to determine if the network is capable of supporting the network speeds desired.
    A. Network specifications
    B. Fiber specifications
    C. Statistical models of the fiber optic cable plant
    D. Loss budgets

Chapter 11

Reflectance Testing

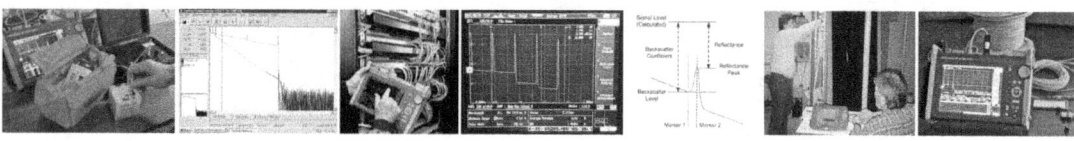

Objectives: From this chapter you should learn:
What is reflectance
How reflectance differs from optical return loss
What causes reflectance
How reflectance is tested

Reflectance And Optical Return Loss

Reflectance or optical return loss (ORL) of a joint between two fibers is the amount of light that is reflected back up the fiber toward the source by the ends of the ferrules of mated connectors or some types of mechanical splices. It was sometimes also called "back reflection" in the past and "Fresnel reflection" from classical optics. Optical return loss is also used to define the total amount of light returned to the source from a long cable plant, including the total of all reflectance from joints and the total backscatter from the entire fiber length.

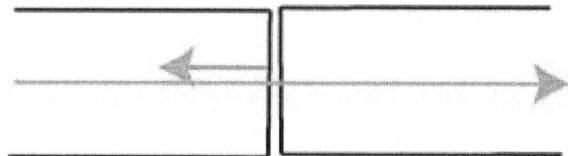

Reflectance is the reflected light from a fiber joint

Causes Of Reflection
Reflections at fiber joints are caused by the light interacting with the change in index of refraction at the interface between the fiber (n=~1.5) and air (n=~1). Reflectance is primarily a problem with connectors but may also occur with mechanical splices, even those that contain an index matching gel to prevent reflectance. Properly made fusion splices should have no reflectance; a reflectance peak indicates incomplete fusion or inclusion of an air bubble or other impurity in the splice.

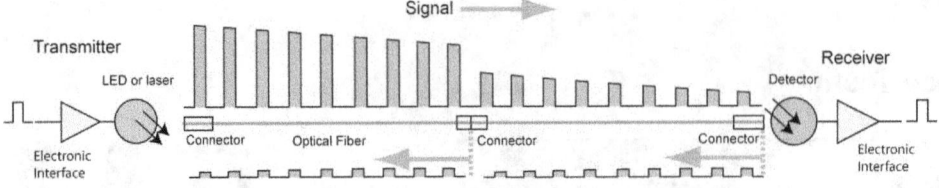

Reflectance is signal reflection from joints between fibers

Reflectance can be a cause of transmission problems in some communications systems. Reflections close to a laser transmitter may cause noise or nonlinearities in the laser. Reflections anywhere in a system may reflect back and forth in a fiber and cause multipath interference or add to the background noise in the system. The effect of reflections can be seen in OTDR traces as "ghosts" as described in Chapter 7.

Minimizing reflectance is necessary to get maximum performance out of high bit rate laser systems and especially AM modulated CATV systems. In multimode systems, reflections are less of a problem but can add to background noise in the fiber at high speeds. Since this is more of a problem with singlemode systems, manufacturers have concentrated on solving the problem for their singlemode components but multimode connectors benefit also since any reduction in reflectance also reduces loss.

Reflectance is another component of the connection's loss, representing as much as 0.3 dB loss for flat-polished non-contact or air-gap connectors where the two fibers do not make contact. Such a high loss is the reason that most connectors are physical contact (PC) types. Connectors have different ferrule end finishes to reduce reflectance as well as loss.

Several methods have been used to reduce reflectance, generally by creating a convex polish on the end of the connector ferrule that creates a physical contact (PC) between two connector ferrules. This PC finish reduces reflectance to about -30 to -50 dB depending on the type of polishing technique used. The technique involves polishing the end surface of the fiber to a convex surface to ensure proper fiber contact. On singlemode fiber, PC finishes work even better at a slight angle (8°) to almost totally prevent reflectance (~ -60 dB). These angle-polished connectors are called APC or angled physical contact connectors.

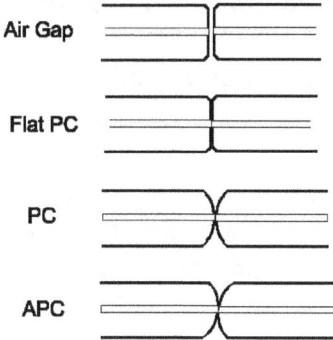

Connector ferrule polish types

Connector Polish Type	Typical Reflectance
Flat with air gap	-20 dB
Physical Contact (PC)	-30 tp -40 dB
Ultra PC	-40 to -50 dB
APC (angled)	-60 dB or higher

Reflectance is defined by the amount of light reflected compared to the power of the light being transmitted down the fiber.

Reflectance (dB) = 10log (P_{refl}/P_{test})

Thus a 1% reflectance is -20 dB, which is about what you get from a flat polished air gap connection, and 1 part per million would be -60 dB, typical of an APC connector.

Optical Return Loss
Optical return loss (ORL) is somewhat confusing because it has several meanings in common usage. For the measurement of reflectance at a fiber joint, optical return loss is stated as the inverse of reflectance or the amount of loss from the signal to the level of the reflectance signal, so -20 dB reflectance would 20 dB optical return loss.

Optical return loss also has another meaning that refers to a slightly different measurement. It can mean measuring the total reflectance from all reflective events in a cable plant under test plus the sum of all the backscatter from the entire length of fiber being tested. While fiber backscatter is small compared to the signal, the summation of the backscatter from a long length of fiber can be significant.

The rationale for this type of ORL testing is somewhat vague, but it appears to be a measure of the noise generated in the fiber by both reflectance and backscatter and that may have some detrimental effect on signal transmission.

The meter and test source method of testing reflectance discussed below will always include both contributions. Most OTDRs can independently measure reflectance at any fiber joint and can also calculate the summation of both for an ORL result.

Measuring Reflectance

There are two ways to measure reflectance. One method uses a test source and optical power meter with some accessories that can be packaged as an instrument called an optical continuous wave reflectometer (OCWR) or return loss test set. The other method uses an optical time domain reflectometer (OTDR). The OTDR method is described in Chapter 7 and below.

The different test methods are used for different purposes. The OCWR method is used primarily for testing connectors on patchcords or short cables, for example in factories making cables. The OTDR method is generally used on installed cable plants where it is important to find and eliminate sources of reflection.

Neither test method is particularly accurate; typically there is about ±1 to 2 dB uncertainty at best. However test equipment manufacturers often have readouts with a resolution of 0.01 dB. This readout resolution often confuses people who think that the resolution of the instrument is the accuracy of the measurement. The problem with measurement of reflectance is the large dynamic range of the measurement and the need to set a reference at very low optical power.

Measuring over a 40 to 60 dB range is always challenging. Reflectance testing adds another problem, how to minimize the errors from other reflecting parts of the cable being tested or even fiber backscatter on longer fibers, especially when setting a reference. Furthermore, reflectance is sensitive to polarization effects that may not be controllable in all test conditions.

OCWR Testing
Below is a diagram of a typical setup for reflectance or return loss testing using a source and power meter per industry standards (TIA FOTP-107 or IEC 61300-3-6). This test is typically used for terminated patch cords when the loss and reflectance values are required specifications for the product as

manufactured.

This test method sends light from a test source through a 2 by 1 optical coupler (a splitter) to the connection that is being tested. That connection will typically consist of a reference connector that is used to mate to the connector under test. The light reflected from that connection is directed back to the instruments where that light is split by the coupler and measured by the power meter.

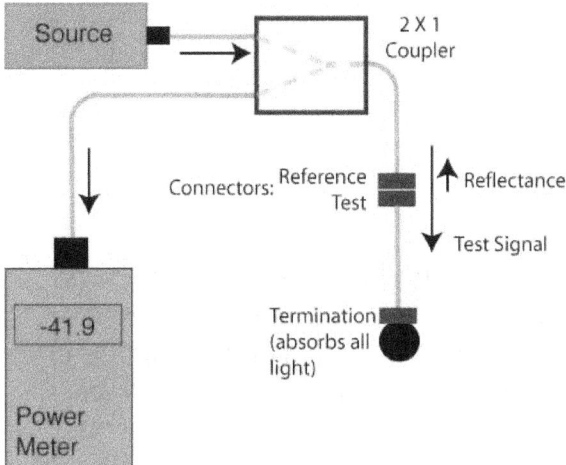

Test setup for OCWR reflectance measurement with terminated connector

Since you are trying to measure the reflectance at the test connection, you need to minimize the reflectance from other components in the system, particularly the connector on the far end of the cable being tested. If that connector were left open to the air, unterminated, it would reflect a large amount of light back that would overwhelm the actual reflectance being tested. This test method requires "terminating" that end of the cable in a way that prevents reflections.

There are two ways often used to "terminate" or reduce the reflectance from the rest of the cable under test. One method is to use an "optical termination" at the end connector, often done by inserting the end of the connector ferrule into an index matching gel or liquid. The index matching gel used for mechanical splices works well, but mineral oil or alcohol works also. After the test, of course, the connector ferrule needs careful cleaning.

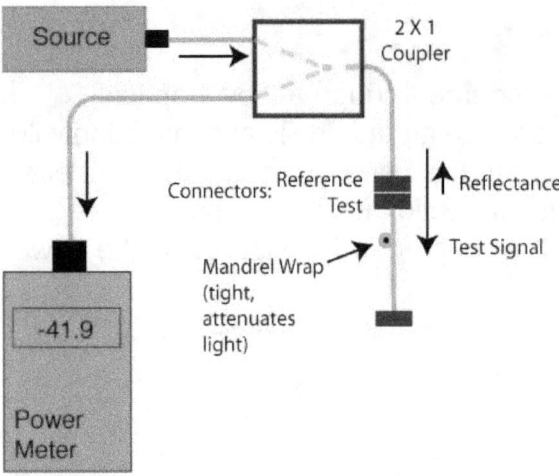

Test setup for OCWR reflectance measurement with mandrel wrap attenuator

The other method that can be used with singlemode fibers is to use a very tight mandrel wrap - about 10 turns on a 6 mm (1/4") mandrel. This should attenuate the signal enough that the reflections from additional components will be minimized. If you are testing a long cable, this method will work best since it attenuates the backscatter contribution from the long length of fiber.

Remember the light will be attenuated in both directions, so if you introduce a 30 dB loss, the total loss will be 60 dB. This method only works on singlemode fiber and it has some drawbacks. It may leave the cable jacket stretched out more than it can recover from, so the cable will always look bad in the area where the mandrel wrap was done. It will not work on patchcords made from "bend-insensitive" fiber, now very common, even for singlemode patchcords.

OCWR Reflectance Tests

Let's look at a complete test and show the process and calculation of the reflectance. First determine the power level of the test source. A test source needs to be fairly powerful if you expect to measure low reflectance, so relatively powerful laser sources are used. Attach a power meter to the reference test connector and measure the power level.

For this example, let's say the test source output is 0 dBm.

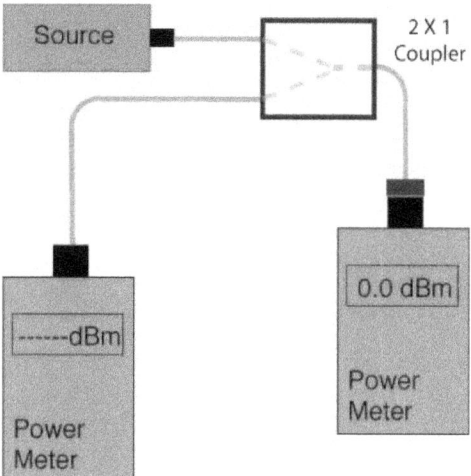

Determining the optical power level of the test source

Next you need to terminate the test connector to absorb all the light to stop any reflections and determine the "noise floor" for the measurement that can include contributions from the coupler and any other devices in the test setup.

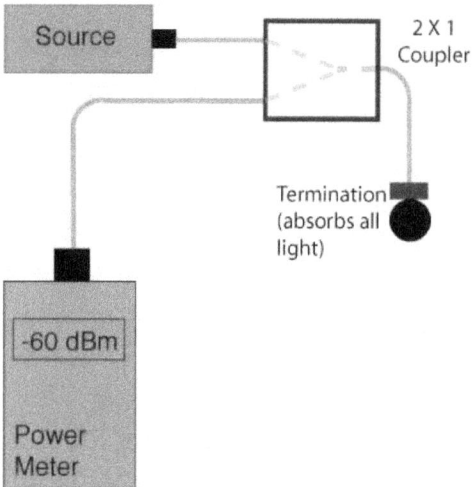

Measuring the noise floor

At -60 dBm, we have a low power level for noise, but it could still be a problem if we tried to check APC connectors.

Now clean the reference connector and attach the cable to test. Terminate the far end using either index matching gel or the mandrel wrap method, which is necessary on a long cable that has not only reflectance events but significant fiber backscatter. This is why this is generally a patchcord test, since testing a connector on a long cable is not possible this way.

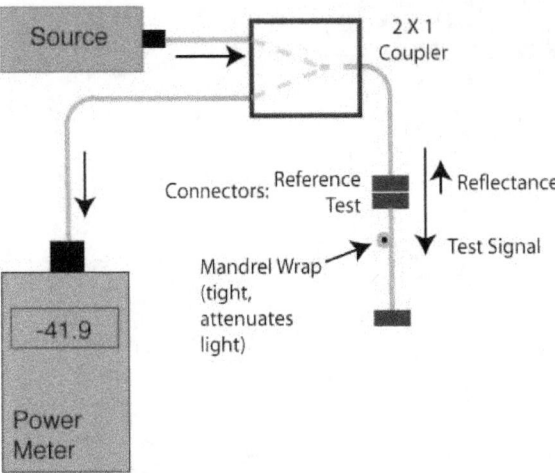

Using a mandrel wrap attenuator to reduce additional reflectance

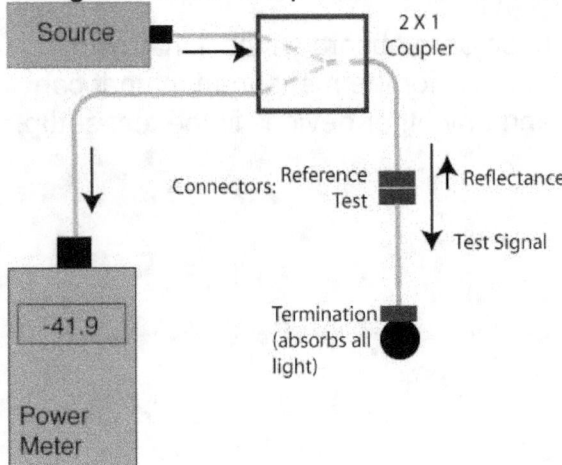

Using a termination on the far end connector on a cable

In order to calculate the reflectance or return loss, you need to know the magnitude of the test signal and the split ratio of the coupler, including the excess loss of the coupler. The coupler will reduce the outgoing signal and reflectance power levels by 3 dB plus some loss due to its inefficiency, typically ~ 3.5 dB. That split ratio should be calibrated beforehand.

The power meter reads -41.9 dBm.

We need to correct the measurement for the coupler split ratio which has split the return signal to the meter in half, so add 3.5 dB to the -41.9 dB measurement.

-41.9 dBm + 3.5 dBm = -38.4 dB

The reflectance is -38.4 dB down from the 0 dBm test signal or a reflectance of -38.4 dB (ORL would be 38.4 dB.) Since the noise floor was -60 dBm, or ~ 1% (20 dB lower) of the reflectance we measured, we can ignore it. If we were measuring an APC connector with -60 dB reflectance, the noise floor would be approximately equal to the reflectance, doubling the power measured which adds 3 dB to the reading. The actual reflectance would be 3 dB lower.

All these references and calculations are made in an OCWR instrument automatically. Some of these instruments use internal references based on a reflection or fiber backscatter. The user should understand the instrument and its operation, as well as its limitations before making measurements. Like the OTDR, the measurement can have dynamic range problems, especially with the extremely low signal from low reflectance connectors or the extremely high signals on open, flat or PC polished connectors.

In use, it is very important to maintain the connector on the instrument. If this connection becomes dirty or scratched, it will directly affect the measurement/

OCWR Testing Long Cables, Measuring ORL
The OCWR technique is really designed to work on patchcords. Since the natural backscatter of the fiber adds to the measured reflectance, longer cable runs will include a significant amount of backscattered light. The total of the reflectance of all events and backscatter is called "optical return loss" (ORL) and is measured as an indicator of the effect of the cable plant on very high speed laser transmitters. OTDRs can also calculate ORL for this same definition. Since longer cables are generally tested with an OTDR anyway, OTDR testing is the preferred method of measuring ORL on long cables.

Reflectance Measurement By OTDR
The OTDR records the amount of light that's returned from both backscatter in the fiber and reflected from a reflective event such as a connector or mechanical splice. The amount of light reflected at a connection is determined by the differences in the index of refraction of the two fibers joined, a function of the composition of the glass in the fiber and any air in the gap between the fibers, common with most terminations and mechanical splices. Most fusion splices will have minimal if any reflectance.

Likewise, APC (angled physical contact) connectors with an angled PC interface have a reflectance of around -60 dB, and probably will not show any real reflectance on an OTDR trace. It may be necessary to refer to cable plant documentation to determine if a low reflectance event seen in a trace is a fusion splice or an APC connector.

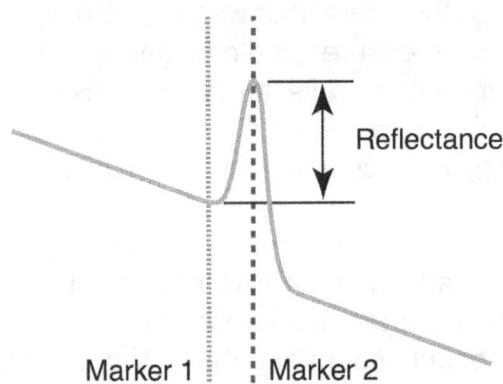

Marker 1 ⋮   ⋮ Marker 2

Reflectance measurement by an OTDR

By choosing the reflectance measurement mode on the OTDR and putting the right cursor on the peak of the reflection and the left cursor just to the left of the reflection at the beginning of the peak, the OTDR will calculate the reflectance of that event.

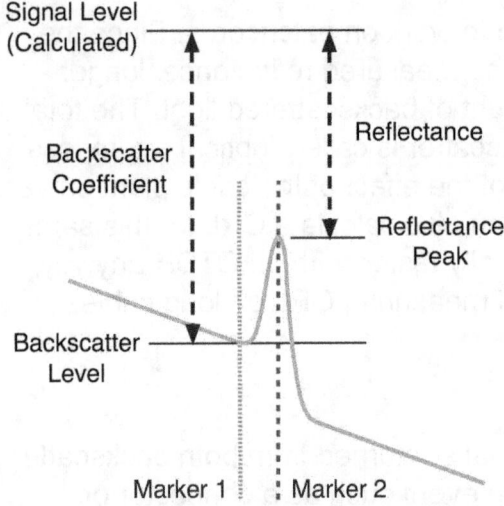

Marker 1 ⋮   ⋮ Marker 2

OTDR calculation of reflectance

This is a complicated calculation involving the baseline noise of the OTDR, backscatter level of the fiber and power at the reflected peak. Since reflectance is defined as a fraction of the amount of power in the test signal, it is necessary to first calculate the test power at that point in the trace – the "0 dB" reference power - and then calculate reflectance. Since the test power cannot be determined directly, it is implied from the backscatter level of the fiber, based on the typical backscatter coefficient of the fiber being tested.

Typical backscatter levels for optical fibers

Wavelength (nm)	Fiber Backscatter Level (1 ns test pulse)
850 (MM)	-67 to -70 dB
1300 (MM)	-74 to -76 dB
1310 (SM)	-77 to -80 dB
1550 (SM)	-81 to -82 dB
1625 (SM)	-82 to -3 dB

The OTDR then measures the backscatter level just before the reflectance peak and uses the test pulse width and the backscatter table to imply test power, then compares that to the power level at the reflectance peak and calculates the reflectance. The indirect way this is calculated makes the method uncertainty high, but provides a simpler method to measure reflectance than the OCWR method.

If the reflectance peak is large and the top of the reflectance peak is flat, it indicates the reflectance signal has probably saturated the OTDR receiver. A saturated peak cannot be used for reflectance measurement since the actual peak height is masked by the dynamic range of the OTDR receiver.

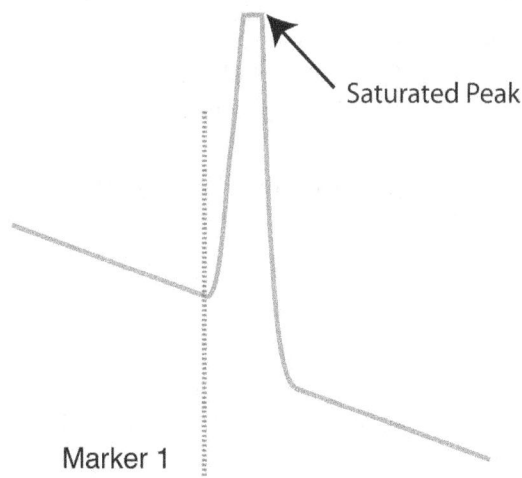

OTDR trace of saturated reflectance peak

Like all reflectance measurements, the OTDR has a fairly high measurement uncertainty. This is caused in part by the difficulty of positioning the cursors properly, especially the right cursor on the peak of the reflectance event. The OTDR has difficulty establishing the levels of the baseline backscatter and the signal peak as both are affected by the test pulse width. And finally, there is the large dynamic range often encountered in measuring reflectance.

While the uncertainty of reflectance measurements is very high, ±several dB, the OTDR technique does have the advantage of showing where reflective events are located so they can be corrected if necessary. Refer to Chapter 9 for more information on testing reflectance with an OTDR.

Optical Return Loss

Optical return loss is related to reflectance. Beside reflectance from individual events, a fiber will have backscatter that the OTDR is using for making measurements. The combined amount of light from reflectance and backscatter is called optical return loss (ORL.) ORL may be a factor in system total noise that affects transmission systems, so OTDRs often can calculate the ORL from the trace. It may be necessary to specify the reporting of ORL but there is no special test setup to measure it.

Chapter Exercises

For these exercises, use either an OCWR or OTDR.
- Test the reflectance of several similar connectors that are both perfectly cleaned and dirty. Compare the results.
- Test the reflectance of several similar connectors that are both new and have been mated many times such as those on old or discarded reference test cables. Compare the results.
- Using an OTDR, compare PC, UPC and APC connectors.
- On an OTDR, compare manual measurements and automatic measurements of reflectance.
- If you have both an OTDR and OCWR, test the same connection and compare results.

Chapter Quiz

1. Reflectance and optical return loss of a fiber joint are the same except for the polarity of the measurement.
>   True
>   False

2. Reflectance and optical return loss of a complete fiber cable plant are the same except for the polarity of the measurement.
>   True
>   False

3. Reflectance is a contributor to the loss of a fiber connection.
   True
   False

4. If a reflectance peak on an OTDR trace is flat-topped, it means you cannot measure reflectance because the OTDR receiver is saturated and the real peak is not measurable.
   True
   False

5. OCWR testing requires _____ the end of the cable properly to get more accurate results.
   A. Terminating
   B. Breaking
   C. Polishing
   D. Cleaning

6. OTDR reflectance measurements depend on knowing the _____ of the fiber.
   A. Attenuation coefficient
   B. Backscatter coefficient
   C. Reflectance coefficient
   D. Length

7. Whether it is being tested by an OCWR or an OTDR, the uncertainty of a reflectance measurement is about _____.
   A. ~ 0.01 – 0.05 dB
   B. ~ 0.1 – 0.2 dB
   C. ~ 1 – 2 dB
   D. ~ > 5 dB

Chapter 12

Testing A Passive Optical Network (PON)

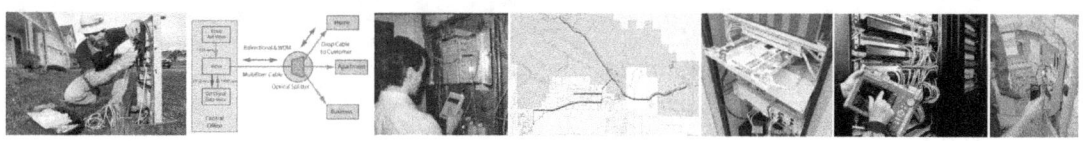

Objectives: From this chapter you should learn:
What is a passive optical network (PON)
Different PON architectures
Insertion loss testing of PONs
OTDR testing of PONs
The differences in upstream and downstream OTDR tests

Passive Optical Networks (PONs)

New network architectures called PONs (passive optical networks) have been developed that allow sharing expensive electronic and electro-optical components by replacing active switches by passive splitters. This network architecture is used for fiber to the home (FTTH) and passive optical LANs (OLANs.) This PON architecture has many advantages. It is less expensive, easy to upgrade and easy to install, but it offers new challenges in testing.

In a PON, a passive splitter takes one signal from an optical line terminal (OLT) and broadcasts it to many users by splitting the signal and sending it to typically as many as 32 optical network terminals (ONT), one at each user downstream. The number of users is somewhat flexible depending on the length of the cable runs and the particular PON standard being used. The splitter adds considerable loss to the links, so one can have a full number of users on short links or fewer users on longer links, since a factor of two in split ratios is equal to 3 dB of optical loss. The electronics uses relatively powerful lasers on singlemode fiber and are set up to tolerate the extra loss caused by splitting the signals.

Upstream the splitter combines inputs from those 32 users and sends it back to the OLT on the same fiber. The passive splitter works like a network switch or hub but is purely optical, using no active components and no power. This PON architecture cuts the cost of the network substantially by sharing one expensive laser with up to 32 homes on the downstream link and only requires an inexpensive laser at each home for the upstream link. Besides saving the cost of an active switch, the PON uses only a single fiber to every user, not two, since signals are sent on the same fiber at different wavelengths upstream and downstream.

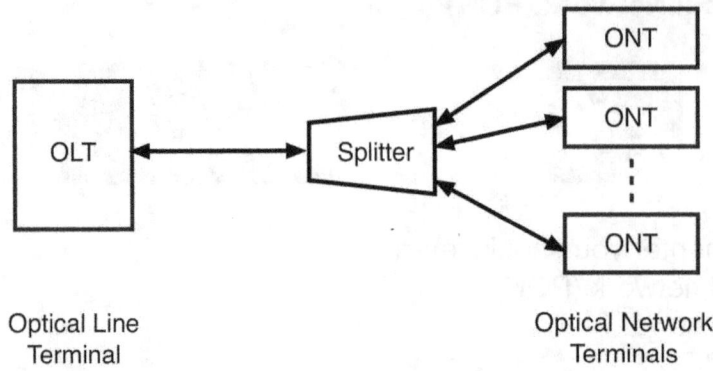

Diagram of a passive optical network (PON)

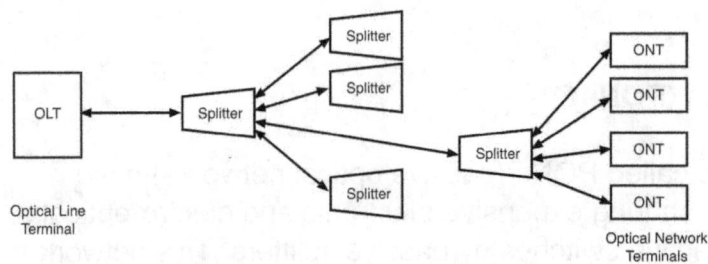

PON with cascaded splitters

The total number of splits of the signal does not have to be done in one splitter. One can cascade splitters to optimize the use of backbone fiber in the system as long as the total numbers of splits does not exceed the maximum. Thus a 32 split can be done as 32 in one splitter, 2 and 16 splits, 4 and 8 splits, for example. The total number of splits is limited by the loss that can be tolerated by the system and the bandwidth of the network.

Needless to say, if you split the light from one fiber into several fibers, the amount of light is divided, so the light in the fibers after splitting is lower by the split ratio. In fact the loss is 3 dB per 2x split (3 dB is a 50% loss) plus a certain amount of "excess loss" caused by the inefficiency of the splitter - loss inside the splitter caused by the mechanism of the splitting.

Splitter losses for common split ratios

Split Ratio	1:2	1:4	1:8	1:16	1:32
Ideal Loss (dB)	3	6	9	12	15
Excess Loss (dB, max)	1	1	2	3	4
Actual Loss (dB)	4	7	11	15	19

The electronics for a PON network are designed to operate on this network architecture. The head end is called an optical line terminal (OLT.) It sends signals downstream with a high power 1490 nm laser that is split and sent to an optical network terminal (ONT) at every user. Each individual user has their data encrypted so only that user can read it, a major security feature that has led PONs to becoming the choice of security conscious users for their LANs.

Upstream, a 1310 nm laser at each user sends data back to the OLT. The electronics are set up to tolerate the extra loss caused by splitting the signals. Early BPON (broadband passive optical network) systems used separate AM TV signals at 1550nm downstream and separate WDM (wavelength division multiplexing) modules or even fiber amplifiers. More modern GPON (gigabit PON) or EPON (Ethernet PON) systems use digital TV and only the two primary wavelengths.

PONs are designed around the splitters so the loss the electronics can tolerate are fairly high. In a typical GPON (gigabit PON) network, the ONT is designed to work with cable plant losses in the range of 13 to 28 dB. This means a cable plant with a 1:32 split will have a splitter with up to 19 dB loss and a margin of 9 dB for the loss of the fiber, splices and connectors in the rest of the cable plant. Knowing those limits is important when calculating a loss budget in the design phase. Having a loss budget is important for comparing to actual test results.

PON Cable Plant Testing

The PON architecture changes the methodology of testing the complete installed cable plant for proper operation. Individual fiber links are tested as usual; it is the PON coupler that creates the difference.

Testing a PON network is similar to other fiber optic testing except the splitter adds complexity. Complete cable plant testing can include some components and installation issues not familiar to the usual fiber tech. PON couplers add high loss and connector reflectance which is not a problem in most systems, can be a problem in short singlemode links typical in PONs. Many FTTx systems use APC (angled PC) connectors to reduce reflectance so reference test cables for both OLTS and OTDR need to have matching connectors.

The actual cable plant can have many different plans, with FTTH and OLANs often quite different. On FTTH networks, you will find longer cables, intermediate splices, splitters placed in many different locations and either spliced or patched in with connectors, and often prefabricated cables using ruggedized outdoor connectors on the user end. OLANs will use shorter cables, patch panels and patchcords for connections. Testing these networks is always simplified by having proper documentation and testing access points.

It is not uncommon for PON networks to add users after installation. That means a link from the ONT will have provision for up to 32 users but fewer users are connected at first. That is a typical way of planning for additional users to be added in the future. That sometimes means that when new users are added and their links need testing, the testing should not disrupt the service to other users.

Many FTTH and OLAN networks are installed using prefabricated cables, called a plug-and-play solution, for the final connection to the user, called the "drop." There is no splicing or termination required so testing is often limited to visual inspection of the connector to confirm it is clean and in good condition.

Testing may be limited to checking power levels at the user's ONT with a calibrated fiber optic power meter or just seeing if the ONT has a "green" connection light. The user's ONT usually has some intelligence that can be accessed from a remote location, allowing a service tech to initiate a loopback test to verify connections at any user. If only one user has a problem, a service tech is then sent there, while if all users are down, the tech is sent to the central office or head end.

As with most fiber optic links, troubleshooting requires knowing the architecture of the system, expected link losses and optical signal levels and typical problems that may be encountered. Inspection of all connectors to insure they are clean and in good condition is a first step in troubleshooting. As always, we emphasize the importance of having documentation on the system before testing and troubleshooting.

Insertion Loss Testing Of PONs

A complete OLT to user ONT link is not a simple run of fiber since it includes the PON splitter and multiple downstream branches. For insertion loss testing with a test source and power meter, the loss of the PON splitter must be included in the loss budget for the link. Since the PON network uses different wavelengths upstream (1310 nm) and downstream (1490 nm), you must measure loss with OLTS at both wavelengths and in the correct direction - similar to how the transmission equipment will use the fiber. Downstream, you would connect the test source and launch cable at the OLT end and test loss at every user location.

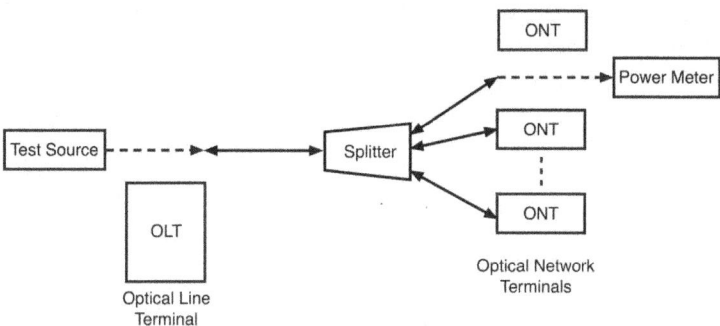

Testing a PON downstream with 1490 nm test source and power meter

Upstream, you would have the meter and receive cable at the OLT location and a 1310 nm test source and launch cable would be taken to every user location. Special FTTx PON OLTS are available that test the proper wavelengths in each direction, simplifying testing logistics.

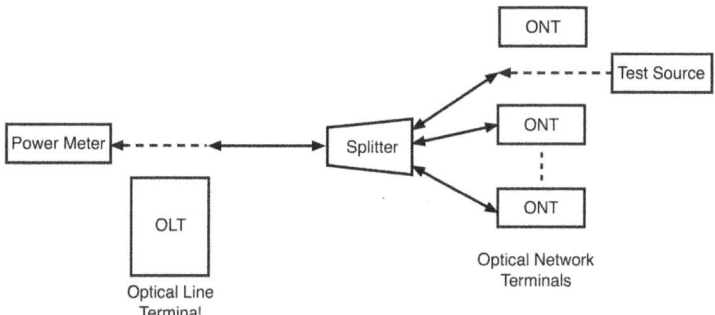

Testing a PON upstream with 1310 nm test source and power meter

OTDR Testing Of PONs

The installer may also need to characterize the PON cable plant with an OTDR, verifying fiber attenuation, termination losses and reflectance and splice quality. The OTDR will also show any bending losses caused during installation. OTDR traces should be filed for future reference.

Since PONs are typically short networks, OTDRs can be used only if the OTDR has very high resolution or the fiber length is adequately long for the OTDR being used. OTDR traces should be done at the wavelengths used in the PON, 1490 nm downstream and 1310 nm upstream. If there are no wavelength division multiplexing couplers in the link, 1550 nm may be substituted for 1490 nm for testing the cable plant.

PON splitters can confuse the OTDR because the traces are more complex downstream than upstream. Here are two traces from an actual PON (FTTH) system taken in two directions.

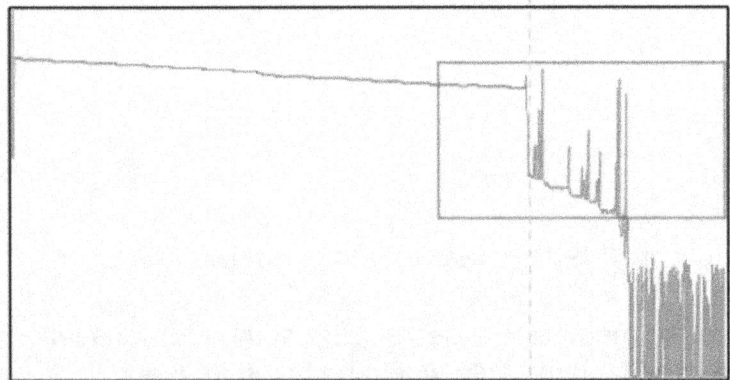

This trace is taken downstream from the OLT to the user:

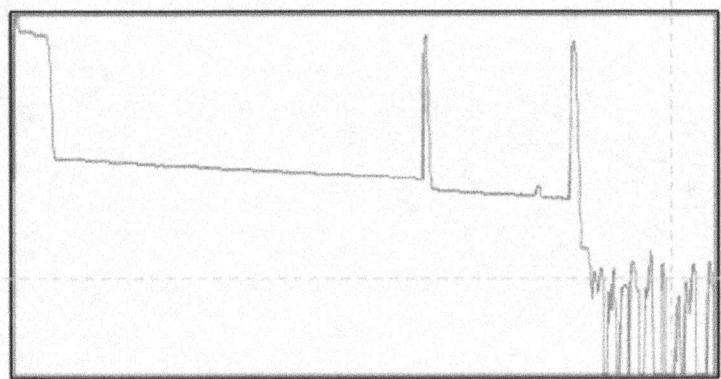

This trace is taken upstream from one user toward the OLT.

In both traces, you can see the large loss of the PON coupler, best seen in the upstream trace at the bottom, on the left side of the trace. On the downstream trace, it is the large loss marked with the dashed marker line, preceding the multiple peaks of the subscriber fibers (outlined in the small box.) The differences in the two traces can be confusing. Below we will show a simpler coupler and explain what you are seeing here.

OTDR Testing Downstream
PON systems create problems for OTDRs. Shooting downstream from the input of a PON splitter at the OLT, the OTDR sees and adds together the backscatter traces from all the fibers connected to the splitter. As a result, it becomes impossible to see detail on individual fibers, and an event (connector, splice or bending loss) cannot be easily assigned to any individual fiber unless the cable plant is carefully documented at installation.

To illustrate this, we will use a schematic of a simple 1X4 splitter system and a schematic of the fiber trace. It's worth noting here that when testing PONs, adding a receive cable to the end of each fiber can make the traces more confusing.

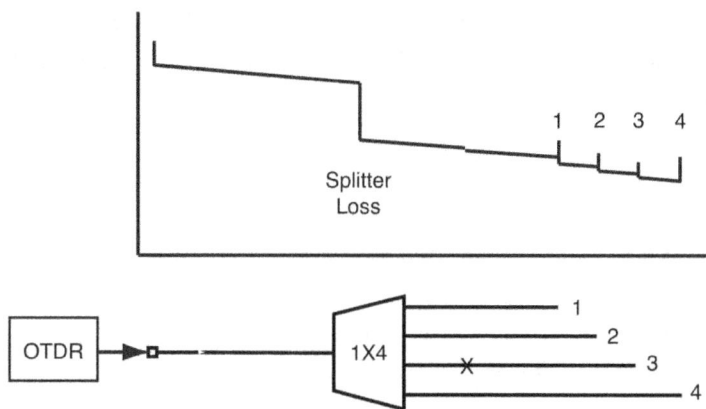

OTDR trace downstream with 1X4 splitter

Consider the "X" shown in the network diagram on fiber number 3. If it is a loss or reflective event, it will show on the OTDR trace (look closely), but the operator would not know if it were in fiber 1, 2, 3 or 4. The only unambiguous part of the OTDR trace shown is the end of fiber 4, the longest fiber, beyond the length of the next longest fiber, #3. This illustration is for only 4 fibers. Think what it would be like with as many as 32 fibers.

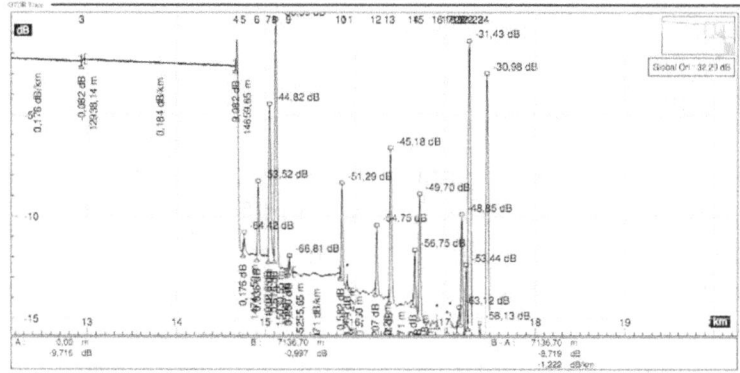

Detail of downstream trace shown above

Here is the actual OTDR analyzed trace shown above, an enlargement of the section in the box. While some newer OTDRs have provision for analyzing PONs, you can see the difficulty of making sense of downstream traces.

It should be noted that FTTH links, because of their short lengths and the use of some high power transmitters, often have APC connectors or fibers prepared to have minimal reflectance. That can make analyzing downstream OTDR traces very difficult when little or no reflective end is available to mark the fiber end and there are 32 fibers in the system.

As a result of the complexity of downstream traces, OTDRs are generally used on PONs from the user end toward the OLT to characterize the fiber path. However the OTDR may also be used from the OLT end or after the splitter to characterize the length of each user fiber link, providing actual fiber length to add to network diagrams for future troubleshooting.

OTDR Testing Upstream
Testing from the user end is easier because you generally test just one fiber. The fiber path will show the events on just one fiber, like the splice (X) shown on fiber 3, and a high loss for the coupler. Here a 1:4 coupler will have 6 dB of splitting loss plus perhaps 1dB excess loss for a total of 7 dB loss.

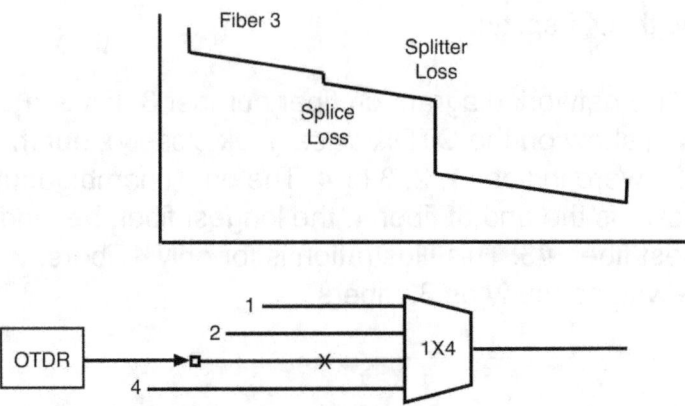

Upstream trace with 1X4 splitter

There is a possible problem with upstream testing. Some systems do not us 1XN splitters, but instead use 2XN splitters. The second fiber allows for testing out of band and having a backup OLT. In this case, there will be some confusion on the 2-fiber side but much less than having many fibers.

Here is a detailed trace from the upstream example above, showing how much simpler the trace is when the other subscriber links are not shown. The fiber on the user end being tested was very short, common with FTTH PONs, so you see the splitter and a long fiber run to the OLT.

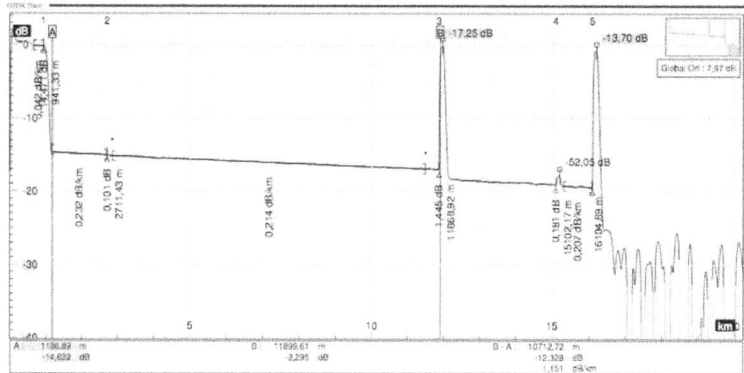

Detail of upstream trace shown above

Other FTTx Testing Issues

Network equipment will be tested as the system is turned on or for troubleshooting. Will the network equipment transmit and receive properly? If the cable plant is installed correctly and tests show loss and reflectance within specifications, it should work. Most FTTx equipment has extensive self-testing capability and that may prove sufficient for most testing. Some PON couplers have a second port on the upstream side (e.g. 2X32 splitter) just for testing or unused downstream splitter output connectors may be useful for testing, especially with OTDRs.

The network equipment should be tested for optical power, comparing the power to the specified power for the PON system. If testing is done while all systems are operating at their respective wavelengths, a power meter with wavelength selective input may be required.

The transmitter output should be within specifications, as should the receiver input, when tested with a calibrated optical power meter set at the proper wavelength(s). Power at the receiver is critical. Too low and the signal-to-noise ratio will be too low; too high and the receiver will saturate. Both conditions will cause transmission errors. High power is not uncommon, so an attenuator may be used in these links to reduce power to acceptable levels.

Data testing with a protocol analyzer is the final test. It will be done using specific protocol testers for the data formats being transmitted. Personnel doing these tests are probably not the same techs that test the cable plant as each often have specific training and test equipment needs. ONTs are generally capable of self testing at the ONT under remote control. This may mean more sophisticated testing is unnecessary for troubleshooting.

Chapter Exercises

Create a loss budget for a PON with 1:32 split where the arrows are fiber links with connectors on each end. The distance from the OLT to the splitter is 1km and each drop is also 1km.

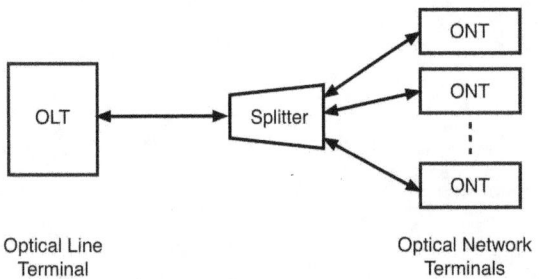

Create a similar loss budget for a PON with 1:32 split, but where the splitters are cascaded - the first splitter is 1X8 and the second is 1X4. As before the fiber links are where the arrows are and fiber links are all 1km with connectors on each end.

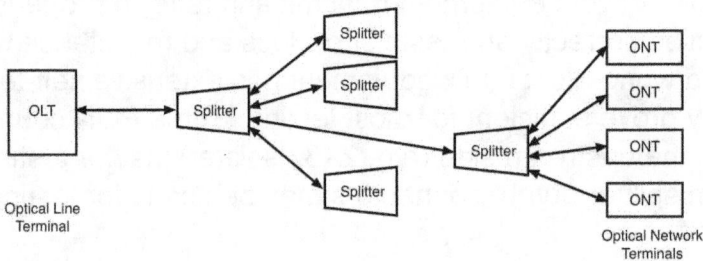

If you have access to an OTDR and a PON cable plant mockup, use that to test it in both directions and view the results. If not, use the FOA OTDR Simulator traces of PONs.

Chapter Quiz

1. As with any fiber optic network, testing is greatly simplified when one has proper documentation for the system.
    True
    False

2. The insertion loss of the link between the CO or head end where the OLT is located and the user at the ONT is measured with an optical loss test set (light source and power meter) and should be measured at all wavelengths used in the system.
    True
    False

3. Connector reflectance is important because of the short links in PONs.
    True
    False

4. When calculating a loss budget for a PON, the splitter or splitters if cascaded should not be included.
    True
    False

5. OLTS testing needs to be done only in either direction on a PON and at either of the wavelengths used in the PON.
    True
    False

6. Each factor of two split in a PON splitter adds _____ dB loss.
    A. 1
    B. 2
    C. 3
    D. 10

7. The differences caused by testing the PON coupler in opposite directions can be confusing when testing with a/an _____.
    A. Optical Time Domain Reflectometer (OTDR)
    B. Optical Loss Test Set (OLTS)
    C. Visual fault locator (VFL)
    D. Optical Power Meter (OPM)

8. The OTDR trace (below) of a passive optical network is taken in the _____ direction.
    A. Downstream
    B. Upstream
    C.

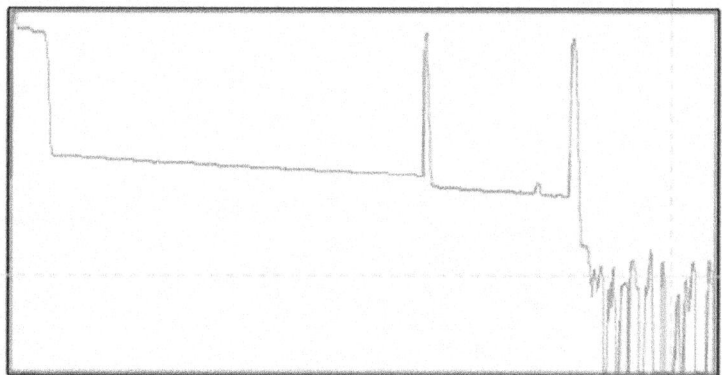

9. In the trace above the large drop near the left side of the trace is causes by
_____.
    A. The loss of the OTDR connection
    B. The loss of the connector on the launch cable
    C. The loss of the PON splitter
    D. A break in the cable

10. Two things which make testing FTTH PON networks different from other
testing are _____ and _____.
    A. PON splitters
    B. WDM (wavelength division multiplexing)
    C. Short cable runs
    D. 1550 nm sources

Chapter 13

Testing And Troubleshooting Checklists

Objectives: From this chapter you should learn:
Why is fiber optic testing required
What tools and test equipment are needed
What can go wrong in installations that testing reveals
How can troubleshooting find problems

Introduction

This chapter is intended as an overview and planning checklist for all managers, engineers and installers on the processes of testing and troubleshooting fiber optic communications systems and cable plants. This document is based on the FOA textbooks, the FOA Online Reference Guide, FOA Technical Bulletins and FOA Standards including the NECA/FOA 301 installation standard. You can download a copy of these standards from the FOA website as a reference.

Note: Standards: FOA "1 Page Standards" have been created to simplify the standard procedures developed into international standards. These standards are generally written by manufacturers for manufacturers, not installers or users. Often these standards contain considerable background material or other explanations aimed at engineers or researchers that make them difficult to interpret. FOA personnel who have worked on standards committees for more than three decades have simplified some these documents that are important to everyday users into quick references for the fiber optic technician.

Why Is Testing Required?

Once a fiber optic cable plant, network, system or link is installed, it needs to be tested for four reasons:
1. To insure the fiber optic cable plant was properly installed to customer specifications or industry standards.
2. To insure the equipment intended for use on the cable plant will operate properly on the cabling
3. To insure the communications equipment is working to specifications
4. To document the cable plant and network for reference in case of future problems

Tools and Test Equipment Needed
The following tools are needed to test and troubleshoot the fiber optic cable plant, system or link properly.

1. Optical Loss Test Set or power meter and test source with optical ratings matching the specifications of the installed system (fiber type and transmitter wavelength and type) and proper connector adapters. An OLTS that merely tests cable plant loss may not include a calibrated power meter needed for testing transmitter and receiver power, so a calibrated power meter and source are a better choice for link or system testing.
2. Reference test cables with proper sized fiber and connectors and compatible mating adapters of known good quality. These do not generally need to be "reference quality" but only in good condition, generally defined as having connector losses of less than 0.3 dB (0.5 dB max.)
3. Visual fiber tracer and/or visual fault locator (VFL)
4. Connector inspection microscope with magnification of 100-200X and fixtures for proper connectors. Video microscopes are recommended.
5. Cleaning supplies intended specifically for the cleaning of fiber optic connectors.
6. OTDR where appropriate, with long launch and receive cables (~100 m for multimode, ~1 km or more for singlemode)

Testing And Troubleshooting The Installed Cable Plant
All fiber cable plants require certain basic tests to insure they were installed correctly and meet expected performance values. These are guidelines for testing and troubleshooting the cable plant itself. The most valuable data one can have for troubleshooting is the installation documentation.

Note: Cleaning Connectors: Before any testing, connectors should be cleaned carefully to ensure that no dirt is present on the end face of the connector ferrule as this will cause high loss and reflectance. Protective caps on connectors, often called "dust caps" – some say that's because they usually contain dust – do not necessarily keep connectors clean. Use cleaning supplies intended for cleaning fiber optic connectors only as other materials my leave residue or cause harm to the connectors.

What Documentation Is Required For The Tests Performed?
Like every step of the fiber optic design, installation and operation processes, it is important to fully document the test and record all relevant data. All tests should be recorded with the following data:

For all tests performed at one time:

1. Date of the test
2. Location of the test
3. Technician performing the test
4. Environmental conditions (temp/humidity/local conditions)
5. Cable plant identification (cable type/fiber type/connector type/length)
6. Type of test (visual inspection, insertion loss, OTDR, CD, PMD, SA)
7. Test equipment used (type, brand, model, serial number, date of last calibration)
8. Wavelength
9. Insertion loss: reference method (1/2/3 cable reference methods)
10. OTDR: manual or auto test, setup parameters

For each individual test
1. Identification of component under test (e.g. fiber #)
2. Test results
3. Note if results are filed electronically

What Can Go Wrong?

There are a number of possible problems with fiber optic cable installations that are caused by installation practice. These include:

1. Damage to the cable during installation caused by improper pulling techniques (such as not pulling the fiber cable by the strength members,) excess pulling tension, tight bends under tension, kinking or even too many bends. Most of these problems will be seen on all fibers in the cable.
2. Damage to the fibers in the cable during cable preparation for splicing or termination. Fibers may be broken or cracked during cable jacket or buffer tube removal or fiber stripping. This may affect all fibers in the cable or buffer tube or just one fiber.
3. High loss splices caused by improper splicing procedures, especially poor cleaving on mechanical splices or improper programming of fusion splicers. Most fusion splicers give feedback on problems if the operator is properly trained. Individual fibers can be damaged when being placed in splice trays or tubes of fibers damaged during placement in splice closures.
4. High loss connectors may be caused by bad processes or damage after termination. Adhesive/polish connectors may have poor end finishes or cracks in the fiber at the end of the ferrule or internally. Prepolished/splice connectors (also called splice-on connectors) are generally high loss due to poor mechanical splicing processes during termination causing high internal loss. All connectors may become dirty or contaminated and should be visually inspected, cleaned and verified by a final visual inspection.

Testing And Troubleshooting Steps For Installed Cable Plants

Before installation, it is advisable to physically inspect all cables as received on the reel for damage and test for continuity using a visual tracer or fault locator. Cables showing signs of damage in shipment may need OTDR testing to determine if the cable itself is damaged. Obviously, no cable showing damage should be installed. Bare fibers can be inspected using an OTDR with launch cable and mechanical splices or bare fiber adapters.

Test Insertion Loss After Installation
1. After installation, splicing (if applicable) and termination, all cables should be tested for insertion loss using a source and meter or OLTS (optical loss test set) according to standards OFSTP-14 for multimode fiber, OFSTP-7 for singlemode fiber. *See FOA Standards for simplified explanations of these standards: http://www.thefoa.org/tech/ref/1pstandards/*
2. Generally cables are tested individually (connector to connector for each terminated section of cable and then a complete concatenated (that means connected together) cable plant is tested "end-to-end", excluding the patchcords that will be used to connect the communications equipment which are tested separately.
3. It is the concatenated cable test that is used to compare to the link power budget and communications equipment power budget to insure proper operation.
4. Insertion loss testing should be done at the wavelength of intended operation if known or at two wavelengths with appropriate sources (850/1300 nm with LEDs for multimode fiber, 1310/1550 nm with lasers for singlemode fiber, 1490 for FTTH.)
5. Unless standards call for bi-directional testing, double-ended testing with both launch and receive cables is adequate.
6. Data on insertion loss of each fiber should be kept for future comparisons if problems arise or restoration becomes necessary. Recording data on a label inside the patch panel or enclosure is common practice.
7. Long cables with splices should be tested with an OTDR to confirm splice quality and detect any problems caused during installation, but insertion loss testing with an OLTS (light source and power meter) is still required to confirm end-to-end loss. Cables with insertion loss near expected values may not also need OTDR testing. Cables tested with an OTDR should have the data kept on file for future needs in restoration. OTDR testing of singlemode cables at 1625nm is sometimes done to find stress losses since the fibers are more sensitive to stress loss at longer wavelength.
8. Long cables intended for very high speed networks require fiber characterization including testing for spectral attenuation, chromatic dispersion and polarization mode dispersion.

Troubleshooting

1. First determine if the problem is with one or all the fibers in the cable. If all fibers are a problem, there is a likelihood of a severe cable installation problem. If all fibers are broken or have higher than expected loss, an OTDR will show the location of the problem on longer cables but premises cables may be too short and need physical inspection of the cable run. If the problem is caused by kinking in the cable or too tight a bend, the cable will have to be repaired or replaced. Generally OSP cables will be spliced as in a restoration unless the cable is a short OSP cable or a premises cable, where it will be replaced.

2. Individual high loss fibers have several potential causes, but bad splices or terminations are the most likely cause for field terminated cables. In some cases, using improper splicing or termination practices will result in high loss for all fibers, just as in kinking or bending losses, not just one fiber.

3. Splices should be tested with an OTDR. High losses may be due to poor splices or bending losses in splice closures. Testing at multiple wavelengths should show higher loss at longer wavelengths on bending losses. Inspect splice closures for improperly routed fibers.

4. Splice loss problems can be pinpointed during OTDR testing. Confirmation with a VFL should be done if the length from the end of the cable is short enough (~2-3 km) where a VFL is usable. The VFL can find high loss splices or cracks in fibers caused by handling problems in the splice tray.

5. Testing of connectors on high loss fibers should start with microscope visual inspection of connector ferrules for proper polish, dirt, contamination, scratches or damage.

6. If dirt appears to be the problem, clean the connectors and retest.

7. If other connector damage is found on visual inspection, re-termination will probably be necessary. Sometimes scratches can be polished out with diamond film by an experienced technician.

8. Prepolished splice connectors with internal splices will generally look OK when inspected with a microscope unless damaged after installation. The most likely cause of loss with these connectors is high splice loss in the internal splice. They can be tested with a visual fault locator coupled into the connector with a short patchcord. High light loss will be seen as an illumination of the connector ferrule or back shell of the conncetor. Some connectors have translucent back shells and can be easily tested with a VFL coupled directly into the connector.

9. If the reason for high loss is not obvious and the connectors are adhesive/polish style, the problem may be a fiber break in the back of the connector. A VFL may help in finding fiber breaks, depending on the connector style and the opacity of the cable jacket.

10. Cables with a fiber or fibers showing very high loss or no light transmission at all should be tested for obvious breaks in the pigtail fiber or cable, generally at the splice or connector, with a visual fault locator or high

resolution OTDR if the cable is of sufficient length
11. High loss links where the excessive loss is only a few dB can be tested like a patchcord with a single-ended test with a source and power meter. When tested in this manner, a high loss connector will show high loss when connected to the launch cable connector but not when connected directly to the power meter detector which picks up all the light from the fiber.

Hints For Troubleshooting
1. Having access to design specifications and installation documentation and specifications will greatly assist troubleshooting.
2. If possible, interview the installer to help uncover processes that may lead to issues in installation, such as pulling methods, lubrication, intermediate pulls, splicing or termination methods (like improper field termination of singlemode which can lead to high loss and reflection even when connectors look OK in a microscope.)

Testing And Troubleshooting Patchcords
Patchcords are short factory-terminated cables usually with standard heat-cured epoxy/polish connectors on each end. They are used to connect equipment to the cable plant and as reference cables for testing insertion loss.

Most patchcord problems are connector problems, caused by damage due to handling or numerous mating cycles when used as reference cables for testing other cables. Connectors may also be damaged by breaking fibers at the back of the connector due to excess stress during handling or by placing other equipment on top of them in enclosures or patch panels.

Testing And Troubleshooting Steps
1. All patchcords, especially those used as reference cables for insertion loss testing, should be tested for insertion loss.
2. Patchcords should be tested with an optical loss test set (optical power meter and source) using single-ended FOTP-171 methods with one reference cable used as a launch cable.
3. This will test the connector mated to the reference cable and the fiber in the patchcord, which is short enough it should have no measurable loss.
4. Since the connector connected to the power meter will not be connected to fiber but presented directly to the detector of the power meter, it effectively has no loss.
5. After testing in one direction, reverse the patchcord and test the other end.
6. In both directions, factory-made patchcords should have a loss of less than 0.5 or whatever performance the user has specified with patchcord

vendors.

7. High loss connectors should be inspected with a microscope for dirt or damage.

8. If other connector damage is found on visual inspection, re-termination will probably be necessary but may not be cost effective, so the patchcord should be replaced. Sometimes scratches can be polished out with diamond film by an experienced technician.

Some optical loss test sets include fiber interfaces on both source and meter ports, so all testing is done double-ended, even if the cable under test is directly connected to an input port. A test set such as this makes reverse testing less effective since reversing test direction may not have any significant effect. Test ports on an OLTS like this should be kept covered when not in use and cleaned periodically. Damaged fibers inside an OLTS will require factory repair.

Testing And Troubleshooting Communications Equipment

After the cable plant has been tested, the communications equipment should be properly connected using matching known-good patchcords. If the cable plant loss is within the loss budget of the equipment (including the loss of the patchcords), the communications link should work properly. If the link does not work, most likely potential problems are the following.

1. Improper connections
2. Cable plant problems
3. Malfunctions of communications equipment

Improper connections

1. The system requires a transmitter be connected to a receiver, of course, so it is important to verify this connection for each link.

2. Even if the cable plant is properly documented, fibers may have been crossed at intermediate connections, so using a visual tracer or visual fault locator will allow quick confirmation of the connection.

3. On long distance links a clip-on fiber identifier can be used to identify individual fibers. Use a test source set at 2 kHz output to allow the fiber identifier to distinguish between dark fibers, live fibers and the fiber being traced.

Testing The Functioning Of The Communications Equipment:

1. If the equipment has correct electrical power and is connected to the cable plant but not operating properly, begin by checking the optical power at the receiver on one end of the link. Some equipment has signal monitoring built-in and will indicate on the receiver if power and signal are being received.

2. Disconnect the cable at the receiver input and measure power with an optical power meter. *Whenever disconnecting cables for testing, inspect and clean connectors.* Do not look into the fiber end in case it has high optical power that could cause eye damage. Make sure the equipment is trying to transmit a signal. Some equipment has a testing mode to force transmission of a test signal or the equipment may simply keep transmitting to try to complete a connection.
3. If the receiver power is within specifications, the receiver or electronics beyond the link may be the problem. Use equipment diagnostics or consult the manufacturer for assistance.
4. If the receiver power is too high, it may be overloading the receiver and an optical attenuator should be inserted at the receiver end to reduce the power to the proper level.
5. If the receiver power is lower than required by operating specifications, the cause is either low transmitter power or too much loss in the cable plant.
6. To test transmitter power, disconnect the patchcord connecting the transmitter to the cable plant and measure the optical power. If the power is low, there is a problem with the transmitter or patchcord.
7. To determine which is the problem, try testing the transmitter with a known good patchcord. If the power is then within spec, replace the bad patchcord and test the link again.
8. If the transmitter power is low with a known good patchcord, the equipment may need maintenance (cleaning) of the output port or replacement.
9. If the transmitter tests as good but receiver power is low, the problem is probably in the cable plant. First try to switch the communications link to spare fibers to see if that solves the problem. Next test the loss of the suspect fibers in the cable plant with an OLTS to determine if the cable plant loss is excessive.

Cable Plant Problems
1. High loss in the cable plant can be caused by damage after installation and testing. Use a visual tracer or visual fault locator to confirm continuity and an OLTS to test loss. See directions above on testing the loss of the cable plant.
2. If the cable plant is long enough, it can be tested with an OTDR to pinpoint problems.
3. If the cable plant loss is not the problem, there are other possible issues related to connection reflectance or the bandwidth of the cable plant.
4. Multimode cable plants may have bandwidth problems caused by the total dispersion due to both chromatic and modal dispersion. Most links have distance limits that differ according to the type of fiber used (OM2, OM3, OM4 or OM5). Ensure the length is appropriate for the fiber type.

5. Multimode cable plants operating at >1Gb/s with 1300 nm with laser sources may have an improperly installed mode-conditioning patch cord (offset-launch) or none at all.
6. Singlemode links that are both high speed and long distance may suffer from problems caused by chromatic dispersion (CD) or polarization mode dispersion (PMD).
7. Singlemode links may suffer from problems caused by reflections at connectors or mechanical splices.
8. Reflections in singlemode connections or splices near the source may cause nonlinearities in the laser transmitter that distort pulse shapes, causing high bit error rates (BER).
9. Reflections near the receiver or at both ends can cause multiple reflections in the cable that create "optical noise" that causes BER.
10. Reflections can be tested, if the cable plant is long enough, with an OTDR to pinpoint problems. Identify the problems by inspecting and cleaning connectors first.
11. Highly reflective connectors or splices should be replaced. Multimode connectors may be repaired by careful re-polishing, preferably using singlemode polishing techniques with diamond polishing film. Singlemode connectors are difficult to repair and should be replaced. Remember most singlemode terminations are made by fusion splicing factory-terminated pigtails onto installed cabling.

Update All Documentation
1. It is vitally important to keep all documentation on the cable plant up to date.
   After completing tests, troubleshooting and repairs, update documentation to reflect the necessary procedures and any changes to the network.
2. If the fix is to switch to spare fibers and suspect fibers are not fixed, note that on documentation to prevent future problems.

Chapter Quiz

1. The only reason to test a fiber optic cable plant is to produce a report so the customer will accept the installation and OK payment
True
False

2. Every cable plant should be tested with an OLTS and an OTDR.
True
False

3. Most fusion splicers give feedback on problems if the operator is properly trained.
True
False

4, Cable plants should be tested for insertion loss and the results compared to _____.
  A. TIA standards
  B. ISO/IEC standards
  C. Manufacturer's specifications
  D. A calculated loss budget

5. _____ in the cable plant near the source may cause nonlinearities in the laser transmitter which distort pulse shapes, causing high bit error rates (BER).
  A. Kinks

  B. High loss

  C. Reflectance

  D. Nonlinearities

6. Long cables intended for very high speed networks require fiber characterization including testing for _____ in addition to visual inspection, OTLS and OTDR testing. (check all that apply)

  A. Spectral attenuation

  B. Chromatic dispersion

  C. Polarization mode dispersion

  D. Exact fiber length

Chapter 14

Metrology and Fiber Optic Measurement Uncertainty

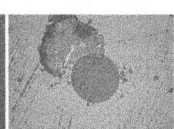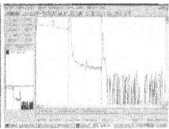

Objectives: From this chapter you should learn:
What is metrology and how does it apply to fiber optics
What is accuracy and precision and how are they related
How do we calculate averages and standard deviation
What contributes to measurement uncertainty for fiber insertion loss testing
What contributes to measurement uncertainty for OTDR testing

Introduction To Metrology - The Science Of Measurements

Metrology is the science of measurements. Metrology covers many issues related to measurements, most of which apply to fiber optics. How does one make measurements? What contributes to the measurement uncertainty? How can one minimize the variables in making a measurement to ensure that on has the least measurement uncertainty? How big is that measurement uncertainty? How accurate is the measurement we make?

Measurement issues affect everyone who is designing, installing or using fiber optics - measurements. We depend on them to find our how well things work, beginning with components at the design level. But we are especially dependent on measurements for verifying the quality and usefulness of installed fiber optic cable plants. We also use measurements to troubleshoot problems.

One common question about measurements is how "accurate" is the measurement. Often the questioner does not really understand what "accuracy" means in technical terms. And the answer is complicated. To provide an answer, we need to explain what a "measurement" is and then what accuracy is. That is the first subject of this technical chapter. Then we will show how this applies to fiber optic measurements for insertion loss and OTDR testing.

Note: The scientists at NIST (National Institute of Standards and Technology, formerly the US National Bureau of Standards) used to correct anyone using the term "accuracy," saying the correct term was "measurement uncertainty." Since everybody generally uses the term accuracy, we will also.

What Is A Measurement?
To understand accuracy, one must first understand "measurement." To measure something means to observe its characteristics and compare it to some standard "unit" of measurement. For example, if we measure length with a ruler, we compare the length of the unknown item to the standard lengths marked on the ruler and express the length in the units that the ruler is calibrated to – mm, cm, meters, inches, feet, etc. Likewise we measure time with a stopwatch calibrated in seconds, minutes, hours, etc. Weight is measured with a scale calibrated in grams, kg, ounces or pounds.

In fiber optics, we measure length with an OTDR, optical power with a power meter, insertion loss with a light source and power meter (LSPM or OLTS), loss and reflectance with an OTDR, etc. We have units of measure for each and instruments to measure them calibrated in the units we understand. You can review these units in Chapter 2 on jargon.

What Is Accuracy? What Is Precision?
Often we hear discussions about the accuracy of a measurement or the precision of a measurement. What is the difference? Here are simple definitions of the two terms.

Accuracy: The degree to which the results of measurement conforms to the actual value or a standard.

Precision: Refinement in a measurement shown by repeated measurements producing nearly the same result.

You can have precision in a measurement without it being accurate, but you must have precision for measurements to be accurate.

Metrologists (the name for those who study or practice the science of measurements) talk about two different terms that we often think of as accuracy. There is "accuracy" that refers to the closeness of the measured value to the actual value and "precision" which refers to the variations you get when making measurements. One needs to understand each well to understand measurements.

If we measure something, accuracy is our goal; we want to know what is the actual value of the thing we are measuring. But when we make the measurement, there are uncertainties in the measurement and those uncertainties contribute to the variation in measurements we make which affect the precision of the measurement. Those variations may be systematic or random.

Measuring a book with a ruler

Let's look at a length measurement of the height of an FOA textbook with a ruler. The measurement we make is going to vary according to how we place the ruler next to what we are trying to measure and how we move our head as we look at the point where the ruler measures the length. Each time we make the measurement, the reading we get can be slightly different, affecting the precision.

Precision is affected by two factors, random errors and systematic errors. Random errors would be the variation each time we read the ruler affected by where our eye is positioned with respect to the ruler and the thing we are measuring. Systematic errors could be caused by the positioning of the ruler at the other end of the object we are measuring, so every measurement has an offset because the ruler and the book end are not perfectly aligned.

The measurement depends on the angle of view - here we might measure too short.

And here we might measure too long.

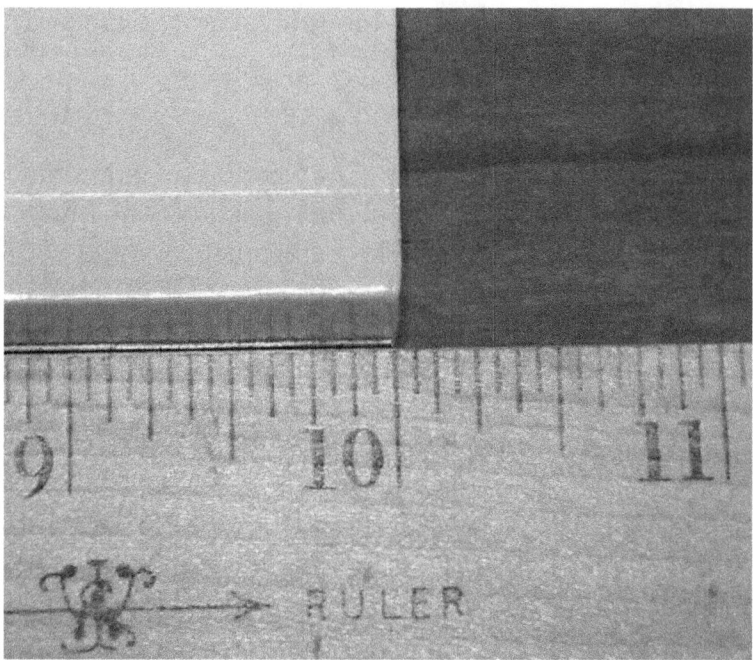

But we get a more accurate measurement if our view is aligned with the end of the book.

The variation of readings based on our viewpoint would be random errors.

If the other end of the ruler is offset from the end of the book, all measurements will be systematically offset from the actual length.

In this case, all measurements would be systematically too long.

The accuracy of the measurement will also depend on the calibration of the measurement tool. If our ruler is not perfectly calibrated, it will cause systematic errors. If the markings on the ruler are hard to read, it can cause random errors. The combination of all the errors means that the length we measure is not exactly the length of what we are measuring. The difference is the inaccuracy of the measurement. We call it the uncertainty of the measurement because we cannot determine what the real factor is exactly.

The Statistics of Measurements

Metrologists look at the statistics of measurements. They set up experiments that minimize the variability of factors that can affect the measurement, make many measurements of the same parameter and do a statistical average of the measurements. They also want to know the precision of the measurement so they calculate the variation, usually calculated as the "standard deviation," the symbol of which is the Greek letter sigma (σ). The standard deviation is an expression of the precision of the measurement, not the accuracy. (See below for more explanation of the math involved.)

The accuracy of the measurement is hard to define because you must define a standard to compare your measurement to – like a ruler for length or a scale for weight. But what do you use for a comparison when you make a fiber optic loss measurement?

One could have a cable plant that was carefully constructed and tested by a standards lab using a source of carefully chosen characteristics like wavelength and multimode fiber mode fill, a fiber optic power meter with extremely good linearity, reference cables chosen to have near perfect fiber geometry (core diameter, ovality, NA, etc.) and terminated carefully to have low loss (~0.1dB or less when tested against each other) and use a lab setup that put no stress on the fibers.

One could then test that cable plant under those careful conditions, get loss values with many measurements so the standard deviation is known and create a "transfer standard." Then the "standard cable plant" could be sent to other labs that would test it under their conditions and get similar data. This is called a "round robin" test and is often done to understand measurement uncertainty. It's done "blind" – that is no lab knows what is the "right answer" so there is no bias in their results.

Then the standards lab gathers all the data and analyzes it, looking for two things: how did the lab's results compare to the "standard test value" of the sample cable plant and how much was the standard deviation of the lab's measurements. The variations in all the labs would also be calculated as a standard deviation, so the data you have includes an average and standard deviation for the round robin.

Now we can start talking about accuracy and precision again. If we assume the standards lab result is the actual measurement because it was made under tightly controlled conditions, the variation of the measurement by each of the other labs is their inaccuracy. And if we average statistically all their variations, it becomes a measure of the precision of the measurements made by labs.

What if we gave that same standard cable plant to a group of random fiber optic installers and asked them to measure it? Then we would have a sample of the real world and the variations could be analyzed as an example of the accuracy and precision of measurements in the real world. Interestingly enough, we are not aware of any program to gather data like that ever attempted.

Setting A Standard For Measurements

We're still not ready to talk about accuracy. To do so, we must have an acceptable "standard" to compare measurements to. For a test like cable plant loss, with so many variables that can cause measurement differences, we are best to tightly specify the variables in the test conditions. In many physical or chemical measurements, it's referred to as "STP" – standard temperature and pressure (atmospheric pressure) - that was always specified for many experiments.

Even standards have some variance – they are not perfect. Standards organizations like the US NIST (National Institute of Standards and Technology) are continually trying to make more precise standards. Length standards, for example, have gone from things like the length of a king's foot to metal bars in controlled conditions to the wavelength of light from a special source.

For our loss testing, it is hard to create a standard that everyone can measure against since a cable plant to send around for comparison would wear out and the loss would get higher as the connectors on the end wore out. Instead we specify test conditions that includes characteristics of the source, meter and reference test cables.

In fact, many if not most standards for fiber optic testing cover all these variables. But, in recognition of the realities of component variations, that still leaves the substantial variability that we have documented in the following sections on testing.

In other words, perfection is unobtainable. We have to learn to live with it.

The Math Of Averages

The average value of a number of measurements is calculated by summing the value of all the measurements and dividing by the number of measurements.

If we make "n" measurements and get results "m1 + m2 + m3 .....mn"

Average (M) = (m1 + m2 + m3 .....+mn) ÷ n

Example: Insertion loss was tested 10 times with the following results (dB):

-10.0, -10.2, -10.3, -9.8, -9.7, -10.0, -10.1, --9.9, -9.9, -10.3

Sum of all 10 measurements ÷ 10 = -100.2 ÷ 10 = -10.02 average

The " standard deviation" (σ) is a measure of how all these measurements vary. It is calculated by calculating the difference between the mean and each measurement, squaring that difference (multiply by itself), adding all those together, dividing by the number of measurements, then taking the square root of that.

Standard deviation (σ) = $\sqrt{\{ [(m1–M)^2 + (m2–M)^2 + (m3–M)^2 + (mn–M)2] / n \}}$

In our example above, we can calculate the standard deviation (hint, we used a formula in a spreadsheet which usually has functions to calculate averages and standard deviation, making this a much simpler process.)

In our example, the standard deviation is ±0.20 dB

Standard deviation (σ) is widely used statistic because it helps understand a standard distribution (Gaussian) of measurements. Approximately 68% of the values lie within one standard deviation of the mean, 95% of the values lie within two standard deviations of the mean and nearly all (99.73%) of the values lie within three standard deviations of the mean.

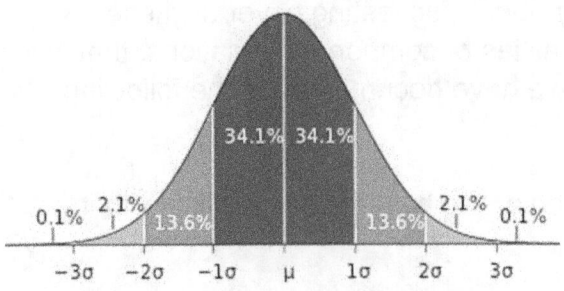

Gaussian distribution of test results

Thus in our example, we would say this measurement was -10.02 dB ±0.20 dB. Based on the test results, we could say the real measurement has a 68% likelihood of being in the range of -9.82 to -10.22 dB (-10.02 dB ±1σ = ±0.20 dB), a 95% likelihood of being in the range of -9.62 to -10.42 dB (2σ) and almost 100% chance of being in the range of -9.42 to -10.62 dB.

The point to remember is that no single measurement is exact - it ha a probability of being correct that is determined from the experimental results.

Metrology In The World Of Fiber Optics
Sometimes a standards lab initiates a program to establish a standard for a new technology. This process was carried out in the mid-1980s for measuring fiber optic power after it was discovered – by your author actually - that power meters from different companies measured the same source but got results that varied by almost 3dB - a factor of 2 times! Measuring power is much simpler than measuring loss since the variables are fewer and easier to control, but that does not mean that creating a standard was simple.

To create a standard for optical power for fiber optics, it was necessary to create a program traceable to NIST standards for optical power. The program calibrated a fiber optic power meter (a highly stable lab model) to the NIST primary standard for optical power using a well defined test source, then sent that meter around to manufacturers to calibrate their working standards for use in manufacturing. That working standard was the meter they used to calibrate the power meters they sold to customers. This brought the uncertainty of power measurements down from 50% to <5%, a major accomplishment. See Chapter 7 for more details on how power meters are calibrated.

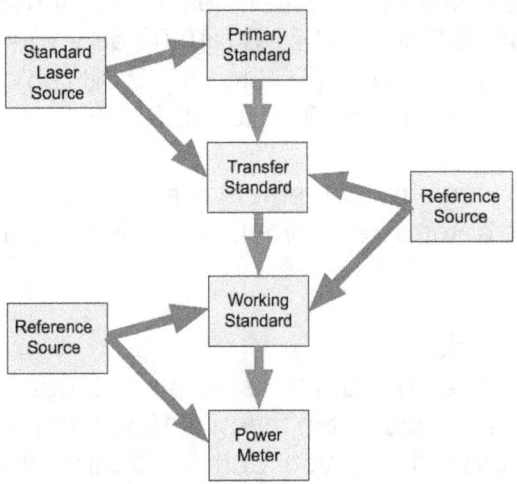

Calibration process for fiber optic power meters

Such a program was once proposed to create a calibration for loss, but the program never got off the ground for lack of agreement on what would constitute a "standard" for loss measurements. Was it a cable plant, a calibrated attenuator or something entirely different? Was one calibration standard applicable to both insertion loss and OTDR measurements? No one could agree. That is unfortunate, since having many instruments measure the same cable plant as part of developing a standard could have provided lots of useful data on the precision and correlation of OLTS and OTDR measurements, it not the accuracy.

In fiber optics, when we do insertion loss testing, we define the units of loss, dB, specify the test equipment characteristics and have each measurement begin with a "0 dB" calibration. As a result, fiber optic loss measurements in the real world are limited in their "accuracy" and precision by how well the equipment and person doing the work meet expected requirements.

Thankfully, the cable plants that are the most critical - like the <2dB cable plants for MM fiber at 850nm for 10G+ transmission - will have the lowest variability because of the shorter lengths of fiber, fewer connectors and subsequently lower loss. And most long, high loss links will be designed to have more loss margin.

Let's examine the two most common fiber optic measurements, insertion loss of a fiber optic cable plant and OTDR testing. As with any testing, to reduce measurement uncertainty, it is important to consider and as, far as possible, control test conditions. It is important to understand all the contributions to measurement uncertainty and how their effects can be minimized. Then we can analyze the sources of error and estimate their uncertainty.

Insertion Loss Measurement Of Multimode Fiber

Insertion loss if the most common measurement in fiber optics so it should be carefully considered, and for the most part it is. The original standards for measuring insertion loss were written in the mid to late 1980s. Even then, most of the potential errors were understood, but over the time that they have been in use, our understanding about what causes measurement uncertainty has become more in depth and our ability to control test conditions improved.

To illustrate the metrology issues, we will focus on insertion loss of multimode fiber optic cable plants that have the most factors affecting measurement uncertainty.

Insertion Loss Error Sources
The table below is a comprehensive - but we cannot guarantee it's complete - listing the sources of error when testing insertion loss of multimode fiber, one of the testing process with the most sources of error. The number of potential sources of error and the magnitude of the errors can be intimidating, but in the real world, many of the errors cancel out and/or are not controllable by the user. Those factors that are controllable will be discussed below with recommendations for minimizing their effects on the measurement accuracy.

In the table we define errors as "systematic" or "random." Systematic errors are errors that affect each measurement in the same way, like using the wrong size fiber for a reference cable or incorrectly controlling modal distribution on the launch cable. Random errors are errors that can be different each time a measurement is made, like dirt on a connection or stress on a launch or receive cable. It is important to understand these types of errors and how to find and minimize them.

Insertion Loss Error Sources – Multimode Fiber

Error Origin	Description	Systematic (S) or Random (R)	Estimated Contribution To Uncertainty (+/-)	Comment
Test Source	LED wavelength and spectral width	S	0-0.2 dB depending on length	Standards allow source wavelength of 820-880nm and fiber has significant change in that range
	Multimode modal distribution	S	0-0.2 dB depending on length and number of connectors or splices	Higher order modes have higher loss so controlling modal distribution – e.g. mandrel wrap is important
	Stability	R	?	Can be caused by instrument design, ambient temperature or battery condition
Launch cable	Regular or BI fiber	S	0-0.2 dB depending on modal distribution	BI fiber acts like fiber with larger core diameter and higher NA
	Fiber core diameter, ovality	S	0-0.2 dB depending on modal distribution	Affects connection loss
	Fiber NA	S	0.1 dB depending on modal distribution	Affects connection loss
	Connector quality	S	0.x dB	Affects connection loss
	Cleanliness	R	0.x dB	Affects connection loss
Receive cable	Regular or BI fiber	S	0-0.2 dB depending on modal distribution	BI fiber acts like fiber with larger core diameter and higher NA

	Fiber core diameter, ovality	S	0-0.2 dB depending on modal distribution	Affects connection loss
	Connector quality	S	0.x dB	Affects connection loss
	Cleanliness	R	0.x dB	Affects connection loss
Meter	Coupling to connector	R	0.x dB	Large detector meters have minimal variation but fiber coupling on OLTS has usual fiber variation issues
	Linearity	S	?	Means a reading of 3dB might actually be 3.1dB
	Stability	R	?	Can be caused by instrument design, ambient temperature or battery condition
Setup	Stress on cables	R	0.x dB	Fiber suffers loss when stressed and can affect connections
Aging	Number of times the connectors on cable plants are connected	R	0-? dB	Loss increases with number of times connections are made as connectors wear
Reference Method	1, 2 or 3 cable reference	S	0.05-0.25 dB	More connectors in the setup for setting the 0dB reference makes measurement different and less precise

Note: Remember that you do not want to add up all these potential errors and assume that would be the potential error for insertion loss measurements. The errors tend to be random and therefore will tend to cancel each other out. It is important to remember that some errors, like multimode fiber modal distribution and the quality/cleanliness of reference connectors are

*controllable by the user and should be given close attention in making
measurements.*

Controlling Test Conditions To Minimize Measurement Uncertainty
Many of these sources of measurement errors are not controllable by the
user. They depend on the manufacturer of the test equipment, optical fiber,
connectors, etc. and their quality control. It is important to use high quality
reference test cables, clean them regularly and test them to ensure they are
still in good condition. Replace them when the loss tested against each other
gets high.

The user should ensure that multimode launch cables have proper mode
power distribution since that can affect the loss measured significantly. Using
a simple mandrel wrap and checking the modal distribution with a HOML test
as described below will greatly reduce measurement errors. For singlemode
fiber, a simple loop mode filter is all that is needed. With all reference cables,
be careful to not stress them during the tests as that can induce loss that will
change the 0 dB reference and or create changes in the modal distribution.

If reasonable precautions are taken, what is the likely accuracy of loss
measurements? Experience has shown that typical measurements have an
uncertainty of approximately ±10% of the measured value in dB. Thus a 2dB
loss has an estimated uncertainty of ±0.2 dB, 10 dB has an uncertainty of ±1
dB and so on.

Modal Distribution In Multimode Fiber
In multimode fibers, some light rays (called "modes") travel straight down the
axis of the fiber while all the others wiggle or bounce back and forth inside the
core. In graded index fiber, the reduction of the index of refraction of the core
as one approaches the cladding causes the higher order modes to follow a
curved path that is longer than the axial ray (the "zero order mode"), but by
virtue of the lower index of refraction away from the axis, light speeds up as
it approaches the cladding and it takes approximately the same time to travel
through the fiber. Thus the "dispersion" or variations in transit time for various
modes, is minimized and bandwidth of the fiber is maximized.

However, the fact that the higher order modes travel farther in the glass core
means that they have a greater likelihood of being scattered or absorbed,
the two primary causes of attenuation in optical fibers. Therefore, the higher
order modes will have greater attenuation than lower order modes, and a long
length of fiber that was fully filled at the launch end (all modes had the same
power level launched into them) will have a lower amount of power in the
higher order modes at longer distances than will a short length of the same

fiber. In fact as light travels down a long fiber, the mode fill will decrease over the length of the fiber and the attenuation coefficient of the fiber will decrease also.

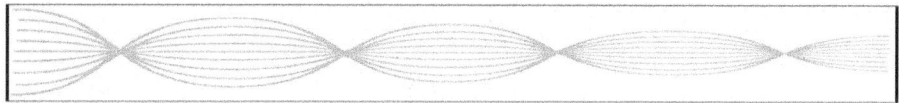

Higher order modes are more attenuated in long lengths of fiber

If we conduct a cutback test on a long length of fiber with an overfilled launch source, perform cutback tests at several lengths of fiber from the far end and calculate the fiber attenuation coefficient, we will get results that look like this.

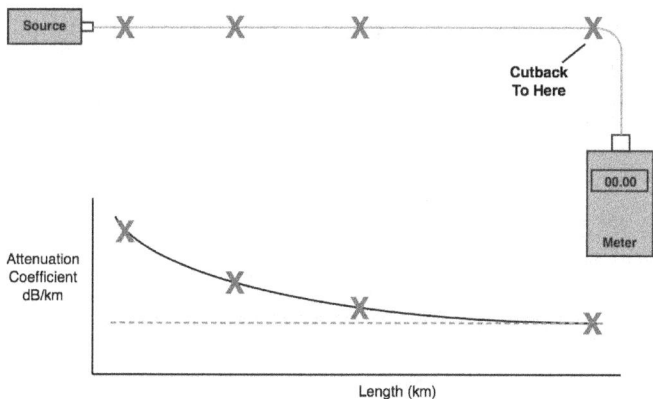

Cutback test on long length of multimode fiber to show transient loss

This change in the measured attenuation coefficient is caused by the change in modal distribution by the attenuation of the higher order modes. This effect is called "transient loss." Transient loss can make a big difference in the measurements one makes of the fiber attenuation coefficient, especially by the cutback method.

The length of fiber not only changes the modal distribution, it changes the attenuation coefficient of the fiber because the higher loss modes are reduced at longer lengths. This effect also reduces the effective core diameter and numerical aperture of the fiber, which affects connection loss since you are effectively transmitting from a smaller diameter fiber into a larger fiber. If you spliced the fibers back together or installed connectors and connected them, the loss would be lower at the longer lengths.

This has always created a problem for specifying fiber attenuation coefficients. That specification depends on the application of the fiber, e.g. short or long lengths. The standard way of quoting the loss of the fiber is to use the long length loss, removing the effects of the higher order modes. That's the dotted line you see in the graph above.

The term "equilibrium modal distribution" (EMD) is used to describe the modal distribution in a long fiber that has lost the majority of the higher order modes. A "long" fiber is one in EMD, while a "short" fiber has all its initially launched higher order modes.

Rather than do a cutback test on a long length of fiber as shown above, the standard method testing is to use a source launch modal distribution that approximates EMD or to condition the modal distribution of the fiber using external means.

Different types of sources launch with different modal distributions. A LED typically has a wider light output that overfills a fiber, while lasers, even VCSELs, underfill the fiber. This also produces conditions in the fiber that affect fiber attenuation measurements and connection measurements.

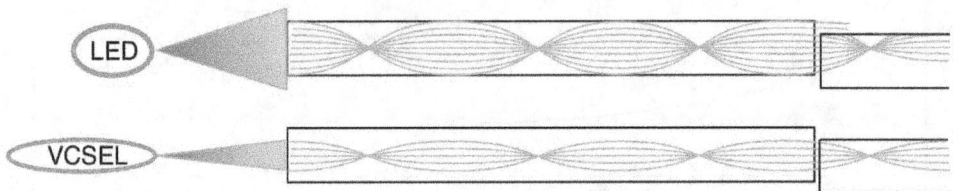

Modal distribution from LED (overfill) and VCSEL (underfill) sources affects fiber attenuation and connector loss

Controlling launch conditions are important to making accurate measurements of multimode fiber. In order to control modal conditions, it is first necessary to be able to measure it.

There have been many methods used to create and measure modal distribution. In early laboratories, modal distribution was measured by the optical power in scans of the fiber output, either in the near field or far field, essentially the same way numerical aperture (NA) was measured. At one point, EMD was simulated by using optics to launch into the fiber with a "70/70" launch, a source with a spot size of 70% of the fiber core diameter and a exit angle of 70% of the fiber specified NA. The use of optics limited that technique to laboratory tests so methods to control modal distribution were developed to allow use in field measurements. The second specification called "mode power distribution" or MPD, was very hard to understand, using templates for the slope of the power output of the fiber. It was never widely used.

The most effective and widely used method of mode conditioning is the mandrel wrap, tightly coiling the launch cable around a specific sized mandrel or rod. The tight bends causes stress on the fiber that produces loss, primarily in the higher order modes. The stress also scrambles lower order modes, making the mode fill more consistent.

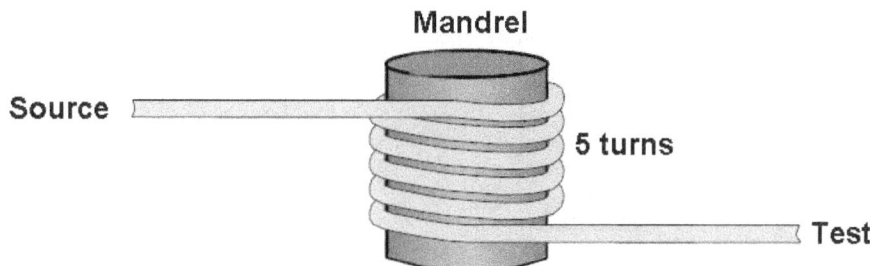

Mandrel wrap mode controller

The mandrel wrap mode controller is based on five turns of a fiber or simplex tight buffer cable around a specified size mandrel. The mandrel size varies according to the cable or fiber type. Below are the TIA standards for the mandrel wrap.

TIA Specified Mandrel Size – Wrap 5 Turns				
Fiber/Cable Type	3mm Jacket	2 or 2.4 mm Jacket	1.6 mm Jacket	900 micron buffered fiber
50/125 micron	22 mm	23 mm	24 mm	25 mm
62.5/125 micron	17 mm	18 mm	19 mm	20 mm

Mandrels are available from test equipment manufacturers or can be made from readily available materials such as a wooden dowel or plastic rod. The 22 mm mandrel is very close to 7/8 inch and 25 mm is 1 inch.

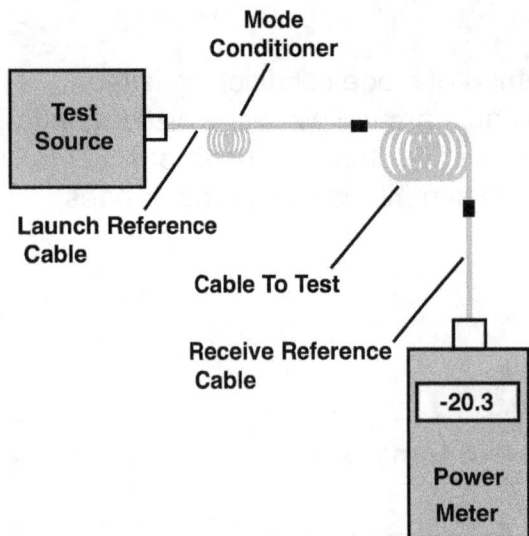

The mandrel wrap mode conditioner should be made on the launch reference cable near the test source

Measuring Source Modal Distribution By Coupled Power Ratio
There was a new method of measuring mode power distribution introduced in the standards with the mandrel wrap mode conditioner around 1990 called "coupled power ratio" or CPR. CPR used a multimode launch reference cable coupled into two different fibers, one a matching multimode fiber and the other a singlemode fiber and measured the difference in loss at the joint between the two.

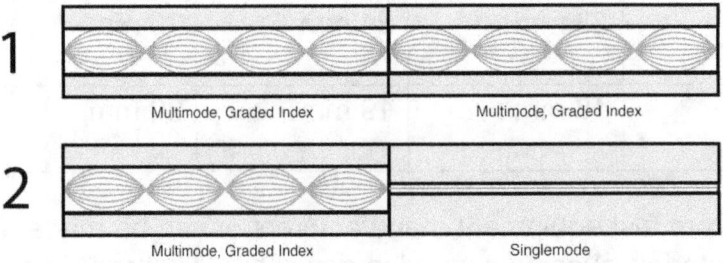

CPR was included in many standards and used for a decade in spite of several problems associated with a misunderstanding of how the test worked. The singlemode fiber required for the test had to be a singlemode fiber for the wavelength of source being tested. That was no problem for 1300 nm testing, but few users knew that a special 850 nm singlemode fiber with core diameter ~4 microns was required at 850 nm and those fibers were hard to find.

Secondly, CPR merely measured the peak power output in the center of the output light pattern compared to the total amount of light. This ignored the important issue of how light was distributed in the higher order modes that was the primary factor in how the launch affected fiber and connector or splice loss.

Measuring Source Modal Distribution By Encircled Flux

As most multimode systems migrated to 850 nm VCSEL sources, a new standard was introduced called encircled flux (EF.) Encircled flux was a theoretical model of modal distribution intended to simulate an ideal VCSEL. It's original use was in modal bandwidth simulations of VCSEL sources that allowed multimode fiber systems to operate at 1 Gb/s and later 10 Gb/s or higher. These systems had very low power budgets, around 2 dB including the penalties for dispersion, so it became more important to reduce the uncertainty of loss measurements.

Encircled flux has been adopted by most insertion loss testing standards for laser optimized fibers today. TIA in the US specifies EF testing for OM3 and OM4 fiber at 850 nm only, considering the fact that legacy systems using older fibers like OM1 or OM2 were tested under prior standards. ISO/IEC standards call for EF testing for all fibers.

Encircled flux is easier to understand than other ways of specifying mode power distribution in an optical fiber. This graph, adapted from the standards with additional notation, defines the standard and illustrates the concept.

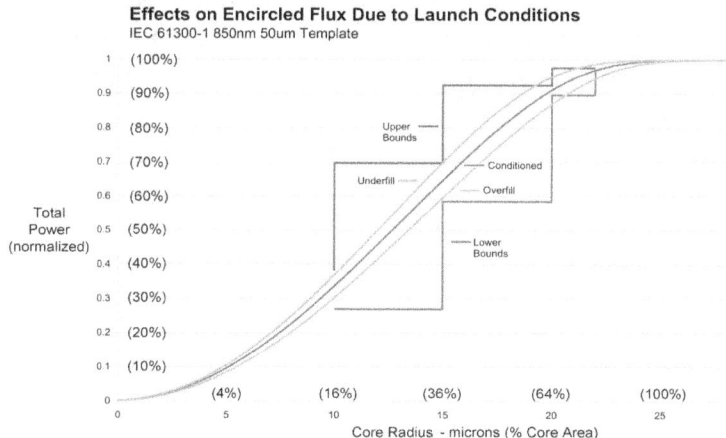

Graph of encircled flux standard for 50/125 micron fiber with annotation

The EF concept is based on the amount of power in the fiber within a specified radius of the core. The graph above is how it is specified, but it is hard to visualize. Basically the power is concentrated in the center of the core, with the intensity of the light decreasing as you get nearer the outer part of the core. Below are two better ways to visualize EF.

Another way to look at EF is to look into the end of the fiber and see how the power is concentrated in the center. The drawing looking into the core of the fiber shows the power levels in shades of gray, with the darker gray at the center and lighter gray at the edges of the core. Note that the outer region of the core has a power intensity only about 1/10th that at the center.

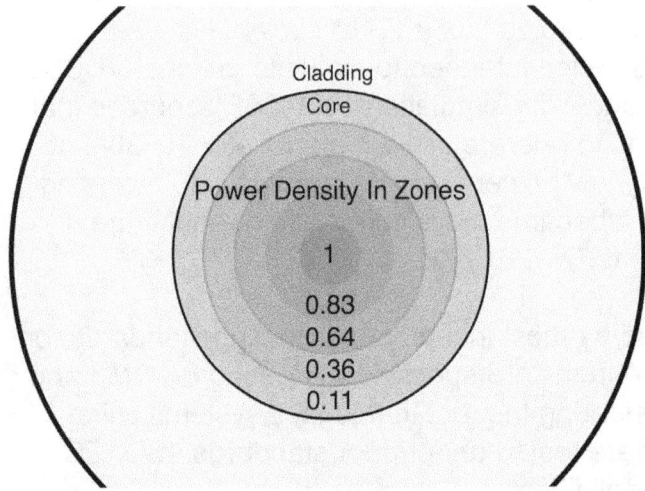

Power in the core of 50/125 fiber meeting encircled flux conditions

Another way to look at EF is with a chart of the power intensity in each region from the center to the outside of the core. As you can see below, the power level is much higher nearer the core. The lower amount of power in the last section is very important for measurements as it is the area that modes have the highest loss in fiber, so fiber attenuation is lower in EF conditions. And this area of the outside of the core is in the region that is most responsible for connector misalignment losses.

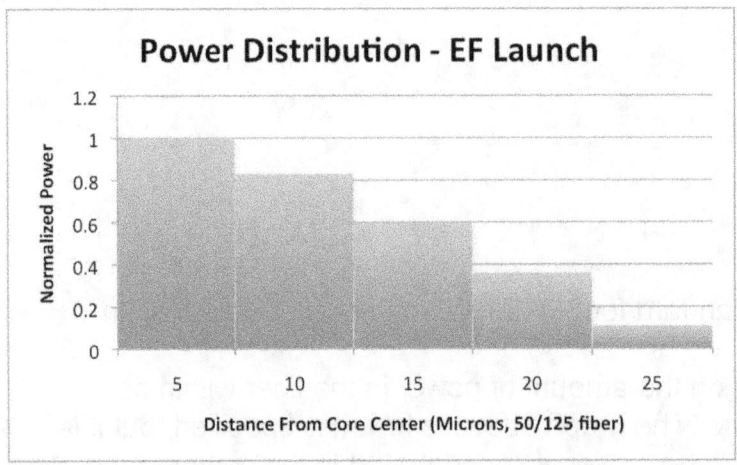

Encircled flux conditions have a major effect on the loss of connectors. The smaller amount of power in the outer layers of the core mean that they have much less influence on misalignment losses in connectors. Here is a graph

showing the relative losses of a connection caused by connector offset with various modal conditioning.

The effect of modal conditioning on connector loss can be very large. Core offset is a major contributor to loss with connectors. With only 2 microns of offset, a fully filled fiber has almost 0.3 dB loss. EMD, which is the lowest mode fill, has little loss, about 0.05 dB. The EF launch, an approximating of a VCSEL source launch, has about a 0.1 dB loss. The graph below shows the effects of mode fill on connector offset loss.

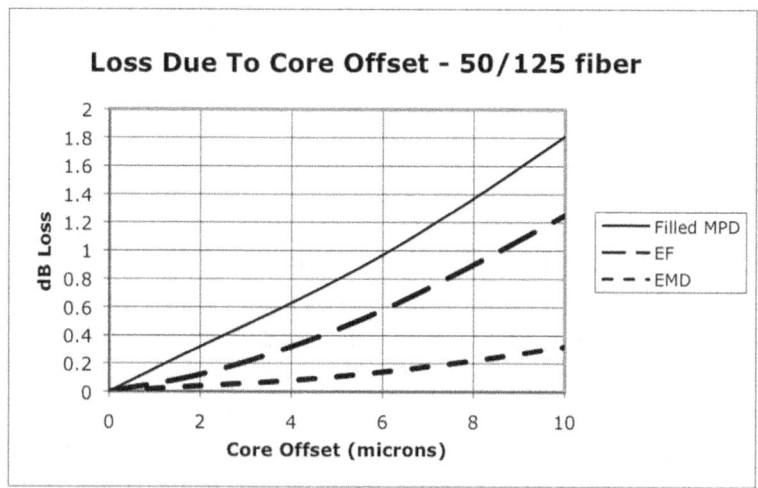

Core offset losses for various modal conditions

Mode fill has similar effects on other parameters that affect connector loss. Fiber diameter variations allowed in standards, for example, allow a 52.5/125 micron fiber to have a core diameter between 47.5 and 52.5 microns. Within that range, mating the largest fiber to the smallest using a fully filled fiber can have almost 1dB loss, while a larger fiber at lower mode fill would only have a few tenths of a dB loss. Similar results come from the allowed core offset in fiber standards. Thankfully, most fibers are much better than these worst-case limits, but they do show the effects of modal distribution in connector loss.

The important things about EF launch conditions for testing are:
- Establishes reproducible modal conditions for testing
- Approximates the typical source (VCSEL) modal conditions
- Causes significantly less typical connector, splice and fiber loss

Thus using EF launch conditions will ensure more relevant, reproducible and *lower* loss measurements. If you are not using EF, measured losses will be higher and sometimes that makes the difference between pass and fail for test results.

Creating And Measuring Encircled Flux Launch Conditions

In order for a standard like EF to be usable, it must be easily created and tested. That was the advantage of the mandrel wrap and CPR. The standards groups that developed EF did not follow through on that aspect of the standard. The initially provided no guidance in the standards on how to create EF conditions and only one company offered mode conditioners that were claimed to create EF conditions.

The standards group also adopted a unique method of testing it. Rather than use optical scanning methods, they chose to simply take a digital photo of the light out of the fiber and analyze the optical power intensity pattern. That was not easy as it requires a digital camera sensitive in the infrared. Instruments of this type are rare and expensive. However manufacturers of other lab equipment using scanning techniques have offered solutions for lab use. In the field, there is no easy solution to measuring EF.

Several years after the standard was approved, an IEC document noted that the mandrel wrap method that had been used for many years essentially met EF conditions. Here is what they said: " *The target weights for 50 μm optical fibre at 850 nm have been studied most extensively. The results were very close to the upper limit of the 10 Gb/s Ethernet limit for transmitters, which means that using it would be conservative; i.e. if the cabling 'passed' when tested using this metric then it would be certain to support 10Gb/s Ethernet. The results were very close to an OFL followed by an 18 mm to 20 mm mandrel with five turns. This is close to what had been defined in some standards as the requirement for testing in premises cabling.*" (IEC document 61282-11 Ed. 1/DTR © IEC (2011)

Higher Order Mode Loss (HOML) Test

The conclusion is that the mandrel wrap will create EF modal conditions, but now it has been determined that it can it be tested. In the TIA, a test was developed and proven to be acceptable in determining if a launch met EF conditions. The test is called "higher order mode loss" or HOML.

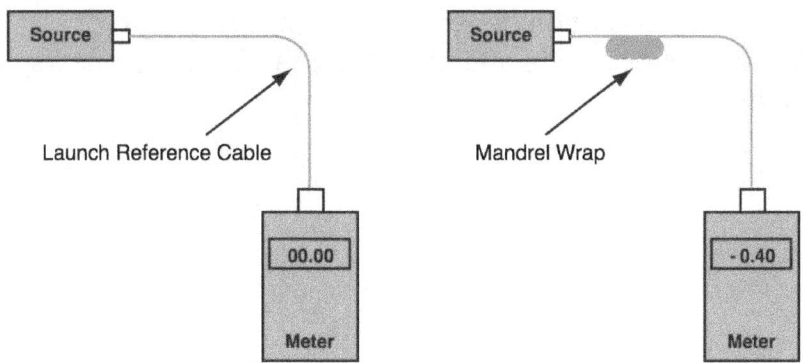

Higher order mode loss (HOML) test for EF launch conditions.

HOML testing is extremely easy. Connect the launch reference cable to a source and measure the output of the reference cable with a power meter. Wrap the launch reference cable around the specified mandrel and measure the output again. There are three options when analyzing the test results.
- If the measured power is reduced by 0.20 to 0.60 dB, the source is essentially EF compliant and ready to use, without the mandrel. Remove the mandrel and make your tests.
- If the HOML is >0.60 dB, leave the mandrel on the reference launch cable and make measurements.
- If the HOML is <0.20 dB, the source has too low a mode fill and should not be used.

Conclusion On Multimode Modal Control
The recommended procedure is to always use the mandrel wrap on the launch cable and use a HOML test to verify the launch conditions. Include these test conditions in the documentation of all testing.

Singlemode Fiber Mode Control
Singlemode fiber only supports a single mode, but when laser sources are coupled into a singlemode fiber, there may be several modes coupled for a short distance. When singlemode fiber is tested with a laser source a mode filter should be included in the launch cable to remove any higher order modes. All that is needed is a small loop in the launch cable near the source with a diameter of 30-60 mm (1.2 to 2.4 inches.) The loop should be taped to make it stable for consistency.

Test Source Wavelength

Optical fiber has different attenuation depending on the wavelength of light being transmitted through it. The attenuation comes from two factors, scattering and absorption.

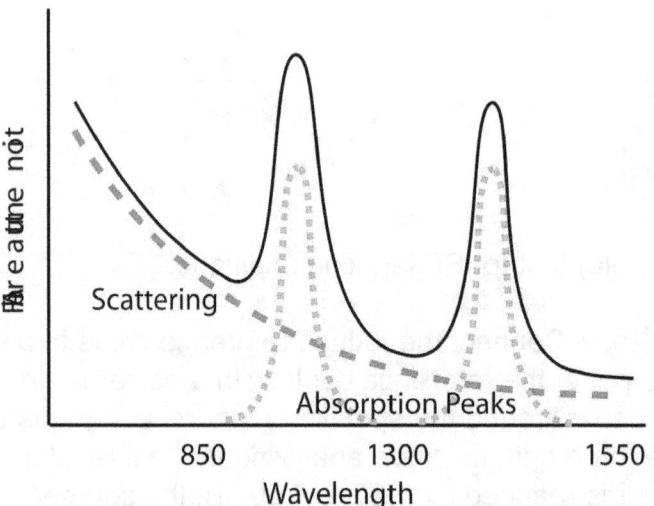

Fiber attenuation is a combination of scattering and absorption

Scattering is a function of the wavelength, decreasing with the fourth power of wavelength (λ^{-4}). Absorption is caused by molecular absorption at specific wavelengths with the most prominent features being OH⁻ absorption (so-called water peaks) at around 950, 1250 and 1380 nm. Newer fibers, called low water peak fibers, reduce the OH- absorption to negligible levels allowing use of the fiber in the wavelength regions of the water bands.

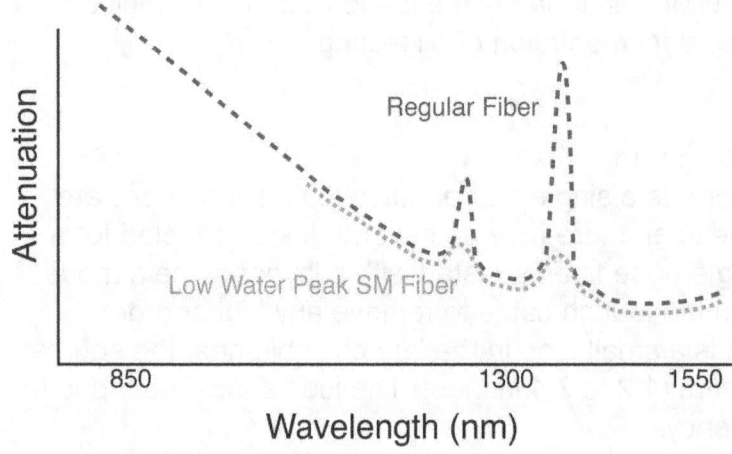

Attenuation of optical fiber

Most fiber optic communications systems avoid the regions of high absorption loss, although the advent of low-water peak fiber has led to the development of coarse wavelength division multiplexing that uses the whole range from 1300 to 1600 nm. For multimode fiber, systems generally use VCSEL sources with a nominal wavelength of 850 nm and a specified range of 840-860 nm. In this range, test sources generally use LEDs which are specified in most standards to have a wavelength of 850 ± 30 nm.

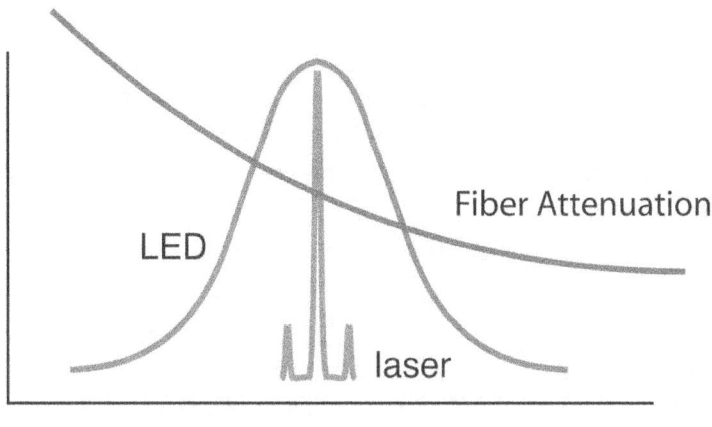

LED and laser source spectral width affect on attenuation measurement

A laser has a very narrow spectral width while a LED source has a very wide spectral width. The laser source tests a fiber at the wavelength of the laser while the LED integrates the measurement over its entire spectral width. Since the fiber attenuation has a dependence on wavelength, the variation in wavelength and spectral width adds uncertainty to loss measurements. The range of wavelengths allowed for 850 nm LED test sources in multimode fibers can cause losses of several tenths of a dB in a few hundred meters of fiber.

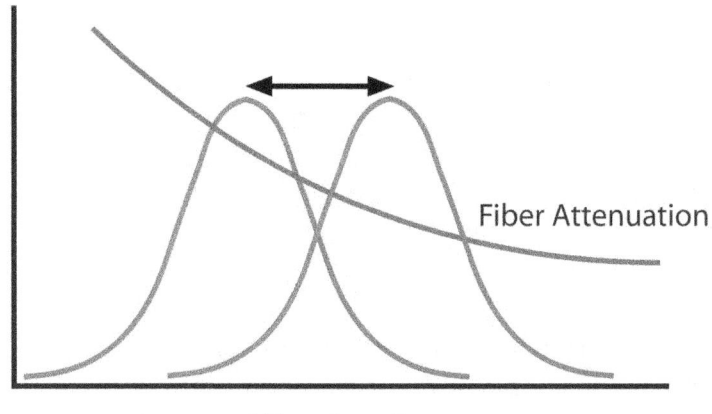

Variation in source wavelength results in difference in fiber loss measurement

OTDR Measurements

Most of the testing questions we get at FOA deal with OTDR measurements. There are several aspects of the questions: How accurate are OTDR measurements? How do they compare to insertion loss tests? How do different OTDRs compare?

Unlike the light source and power meter that measures loss in the same manner as a transmission system works, the OTDR uses a indirect method of measurement based on backscattered light. It also needs longer reference launch and receive cables than insertion loss tests with a light source and power meter because of its need for covering the dead zone and the OTDR's limited resolution.

Sources Of OTDR Measurement Error
While all the variables noted in the insertion loss discussion above affect OTDR measurements in the same manner, the OTDR also has a few unique sources of error. Best known is the difference in backscatter of two fibers at a connector or splice that cause the OTDR to measure a loss that is too high (going from high backscatter to low) or too low, even a gain (going from low backscatter to high.) This error is reduced by measuring in both directions and averaging. More detail is in Chapter 7.

Another major error in OTDR measurements is caused by the placement of the markers that determine where on the trace the measurements are made, even if you use the "least squares" mode to reduce the variations. This creates a random error that affects the precision of the measurement. If you use an "auto test" function, the placement of the markers is done by software in the OTDR and the reproducibility will be a function of the software algorithms.

Which brings up a big issue for OTDRs, software. Software is very much a part of OTDR measurements and a major factor in measurement uncertainty. It controls the setup parameters, controls the test pulse and averages the return backscatter and reflectance signals to create the trace. If you use auto test, software decides how and where to place markers and make measurements.

Software, therefore, plays a big role in the uncertainty of OTDR measurements. The role that software plays in the measurement is hard to evaluate. You could test the same cable plant with many different OTDRs using different software and compare the results, but many of the variables noted above also contribute to the variability of the measurements.

While it is hard to evaluate the actual precision and accuracy of OTDR measurements, our experience is that the precision and accuracy are less than the simpler insertion loss tests.

Sources Of OTDR Measurement Errors

Error Origin	Description	Systematic (S) or Random (R)	Estimated Contribution To Uncertainty (+/-)	Comment
OTDR Instrument Errors	Time base, receiver linearity	S	0.xdB	Affects ability to measure distance or loss, affects length and dB measurements
OTDR setup	Averaging	R	0.xdB	Affects ability to measure distance or loss, affects length and dB measurements
	Marker placement	R	0.xdB	Affects ability to measure distance or loss, affects length and dB measurements
	2 point or LSA measurement	R, S	0.xdB	Varies according to marker placement and length of fiber included
OTDR Autotest	Unknown setup	R, S	0.xdB	Unknown effects due to software selection of parameters
Test Source	Wavelength	S	0-several dB depends on length	Standards allow source wavelength of ~1280-1330nm and fiber has ~0.035 dB/km change in that range

	Multimode Modal distribution	S	0-0.2dB depends on length and number of connectors or splices	Modal distribution in OTDR testing is complicated, with the test signal generally being underfilled and the return signal overfilled.
Launch cable	MM Regular or BI fiber	S	0-0.2dB depends on modal distribution	BI fiber acts like fiber with larger core diameter and higher NA
	Fiber core diameter, ovality SM MFD	S	0-0.2dB depends on modal distribution	Affects connection loss
	MM Fiber NA	S	0.1dB depends on modal distribution	Affects connection loss
	Connector quality	S	0.xdB	Affects connection loss
	Cleanliness	R	0.xdB	Affects connection loss
Receive cable	MM Regular or BI fiber	S	0-0.2dB depends on modal distribution	BI fiber acts like fiber with larger core diameter and higher NA
	Fiber core diameter, ovality	S	0-0.2dB depends on modal distribution	Affects connection loss
	Connector quality	S	0.xdB	Affects connection loss
	Cleanliness	R	0.xdB	Affects connection loss
Cable Under Test	Backscatter coefficient variations	R	0.xdB	May vary from fiber to fiber or along a fiber, causes connection errors
	Index of refraction or group velocity variation	R	~1%	Vary fiber to fiber, affects length measurement and thereby loss coefficient

Aging	Number of times the connectors on cable plants are connected	R	0-?dB	Loss increases with number of times connections are made as connectors wear

Of course if one added up all the possible uncertainties, it would be several dB and measurements would be practically impossible. However, all setups are not worst case and even so, some variables are going to be positive and some negative, so they tend to cancel each other out.

Modal Conditioning In Multimode Fiber

In discussing insertion loss testing of multimode fiber, we spend most of our time focused on mode power distribution and its effects on fiber attenuation and connection or splice loss. With OTDR testing, modal distribution is not as much a topic of consideration.

Modal conditioning for OTDR testing of multimode fiber is a poorly understood subject. The OTDR launches a test signal with a very powerful laser that has a very restricted modal distribution. But the backscatter signal that provides the information for the OTDR to analyze basically fully fills the modes in the fiber for the return path. Therefore you have two very different analyses going on for each test – underfilled modes on the outgoing signal, overfilled modes on the return signal.

Because of this issue, mode conditioning on the outgoing signal seems to make less sense and some proposed encircled flux mode conditioners for OTDRs are not designed to be usable in a reverse direction. This is just another part of the difference in testing methods between insertion loss and OTDR testing that indicates that the two test methods are unlikely to be really comparable in most test situations.

Directional Fiber Joint Loss Errors Caused By Variations In Backscatter Coefficient

When the OTDR test pulse goes through a joint in the fiber made by a splice or a connector, some light is lost in the joint. The reduced light in the test pulse will reduce the amount of light backscatter and that will be measured by the OTDR and shown as a loss in the trace at that point. That loss can be measured using the techniques described above.

The biggest source of measurement uncertainty that occurs when testing loss at a joint with an OTDR is a variation of the backscatter coefficient, the amount of light from the outgoing test pulse that is scattered back toward the OTDR by the fiber. The OTDR looks at the returning signal and calculates loss based on the amount of light it sees coming back and there is no way the

OTDR can determine if the backscatter coefficient of the fibers is a constant.

The light scattered back to the OTDR for measurement is not a constant for all fibers, in fact it's often slightly different for any two fibers. Backscatter is a function of the attenuation of the fiber and the diameter of the core of multimode fiber or mode field diameter (MFD) in singlemode fiber. Scattering is the major cause of attenuation in optical fiber. Higher attenuation fiber has more attenuation because the composition of the glass in its core scatters more light. In singlemode fiber, a larger mode field diameter (MFD) results in less backscatter and inversely, a smaller MFD will have more backscatter. In multimode fiber, bend-insensitive fiber with its core/cladding design that reflects light back into the core has higher backscatter than non-BI fiber.

When the loss of a splice or connector joining two different fibers is tested in an OTDR, any difference in backscatter from each fiber will cause an error that is dependent on the direction of the test. There are three possibilities at the joint of the two fibers.

If both fibers are identical, such as when splicing a broken fiber back together, the backscattering coefficient will be the same on both sides of the joint, so the OTDR will measure the actual splice loss in each direction.

If the fiber nearer to the OTDR has higher backscatter than the one after the connection, the amount of backscattered light will go down after the joint, so the measured loss on the OTDR will include the actual loss in that direction plus a loss error caused by the lower backscatter level, making the displayed loss greater than it actually is.

If the fiber nearer the OTDR has a low backscatter level and the fiber after the joint a high backscatter fiber, the backscatter level goes up, making the measured loss less than the actual loss in that direction. If the change in backscatter level is greater than the splice loss, the joint will be a "gainer", what looks like a gain in the fiber at the joint not a loss, a major confusion to new OTDR users.

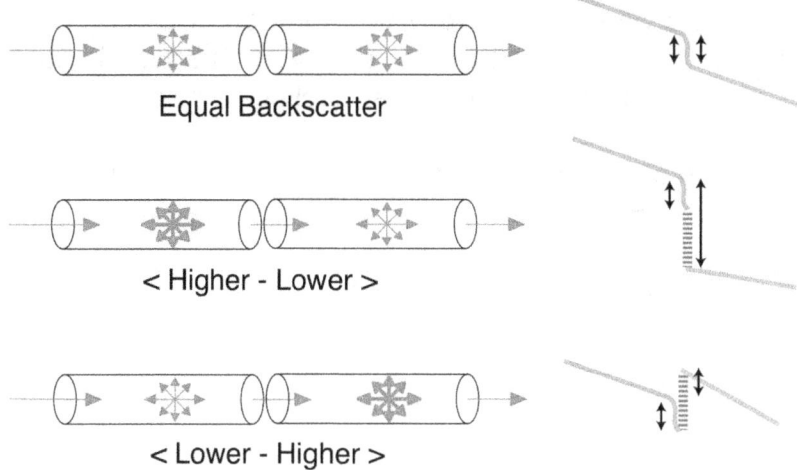

OTDR measurement of a joint between two fibers.

In the drawing of the traces, the wide dashed line in the OTDR trace of the splice represents the difference in backscatter coefficient for the two fibers. That amount of change can be significant. For singlemode fiber, the tolerance in mode field diameter allowed by standards can create a maximum difference in backscatter of almost 1 dB, although more typical production variations will lead to variations less than half that, about 0.4 dB. Even so, the average fusion splice is typically less than 0.1 dB, so one would expect to see a lot of errors. And in fact, you do. Techs report that around 1/3 of all splices will show a gain in one direction, so you know the loss is going to be in error, showing too high on the trace in the other direction.

Here are two ways you may see gainers. First is a simple splice gain at the event in the OTDR trace.

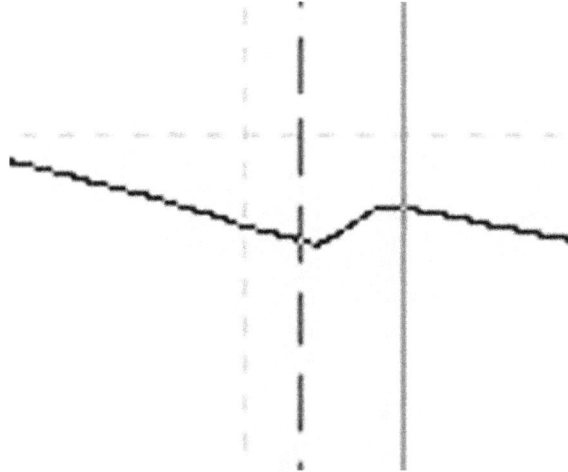

A real "gainer" - a splice 35 km away in an installed fiber link

Another way gainers show up in an OTDR trace is caused by a high backscatter fiber spliced into a very long concatenated cable. In this case, event 1 appears to be a fiber with smaller mode field diameter was spliced into a link between two fibers of larger mode field diameter. At the end nearer the OTDR, the splice shows a gainer due to higher backscatter from the smaller MFD fiber, while at the other end it shows a larger splice loss as the smaller MFD fiber is spliced to a larger MFD fiber. The difference is not small, in this case it is more than 1 dB.

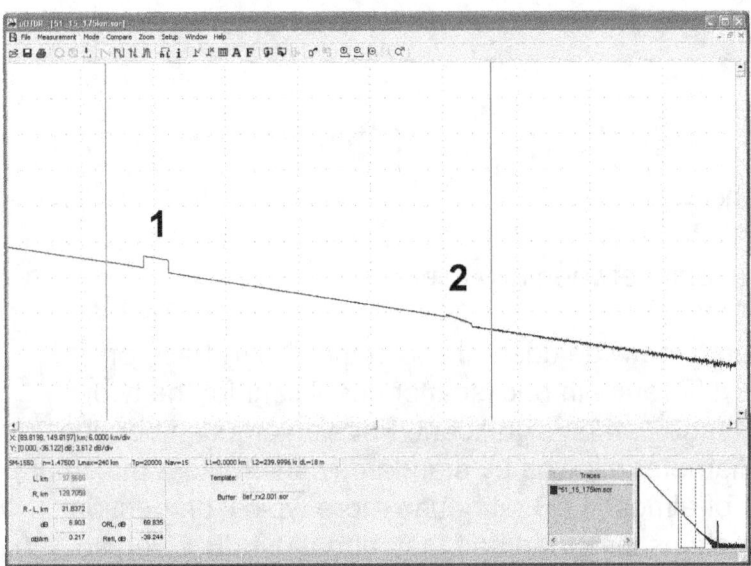

Trace of long concatenated fiber link with gainers

Event 1 shows another important issue about gainers - in the opposite change in backscatter a gainer becomes a big loser - the same difference in backscatter coefficient that causes the gainer in one direction causes a higher loss in the other direction, which can cause considerable problems in evaluating splices if you only test in one direction.

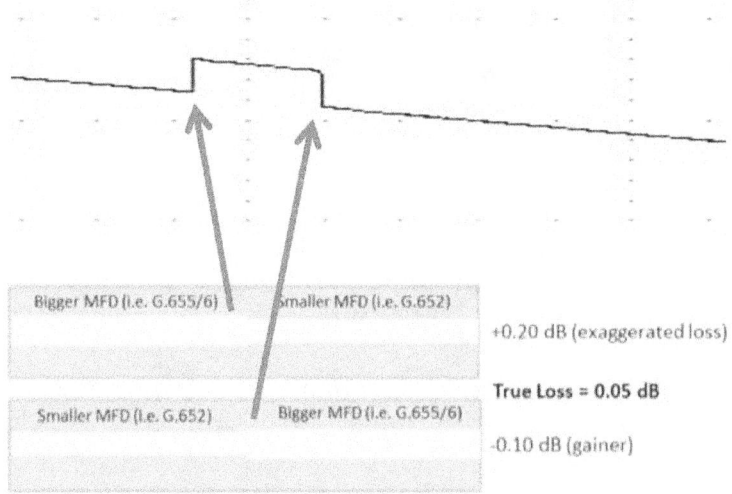

Trace of smaller MFD fiber spliced to larger MFD fibers

Here is an example of what happens when you splice a smaller MFD fiber with higher backscatter to two larger MFD fibers with smaller backscatter. When you see a trace like this, with a section of fiber elevated above other fibers (or below the other fiber traces in the opposite situation), you know it's a fiber mismatch.

In the higher loss direction, the loss shown by the OTDR will always be the actual loss in that direction plus the difference in backscatter coefficient, so even if the loss from the actual splice is very low, the measured loss can be very high. This can cause problems if you try redoing the splice trying to correct the problem, because the OTDR measurement will always be wrong and always show the loss as high no matter how good the splice.

You can see another variation with event #2. Note how the attenuation slope of the fiber is much higher in the fiber in this segment? The fiber in event 1 had an attenuation coefficient similar to the other fibers in the link, so the difference was probably just a MFD variation. Here the fiber has much higher attenuation, so it may be a core composition problem or a cabling problem that put stress on the fiber. If it is a cabling problem, which could be from too much stress on the cable in installation, it is in addition to the variation in backscatter that causes the initial gainer.

The traces below show that this issue is not limited to singlemode fiber. These traces were made as part of an experiment to see what happens when regular and bend-insensitive multimode fiber are joined. As you can see the BIMMF has much higher backscatter and creates a gainer. This is caused by the ring of lower index glass in the cladding that reflects light back to the core if it is lost due to stress.

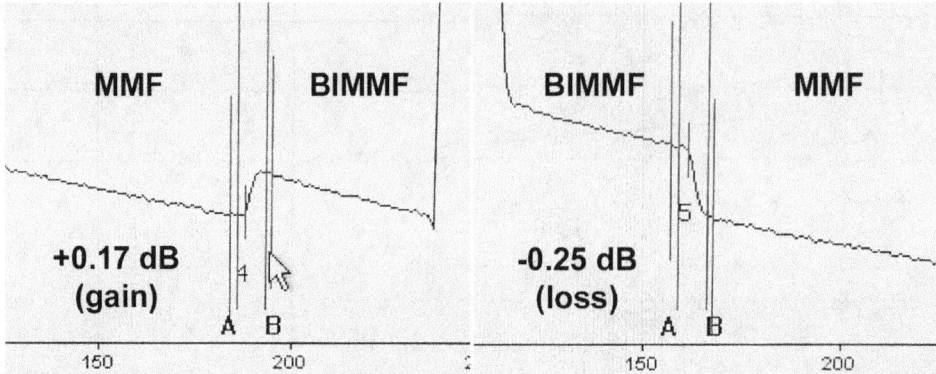

Bidirectional traces of multimode fiber spliced to bend insensitive multimode fiber

There Are Also Real Directional Loss Differences At Joints

Besides the directional loss differences caused by differences in backscatter, there are real directional differences in the loss of many joints. These may be caused by differences in the fiber designs or manufacturing variations.

Within singlemode fibers, there are regular SM fiber, large mode field diameter (MFD) fibers, dispersion shifted fibers and bend-insensitive fibers, plus manufacturing tolerances that can cause fiber differences. Some of these fibers have very different fiber index profiles. Mixing these fiber types often occurs because of the standard practice of splicing regular singlemode fiber pigtails on cables with fibers of any type.

Multimode fiber has manufacturing variations in core size among each fiber type, variations in index profiles of various bandwidth grades and bend-insensitive fiber variations as shown in the traces above.

These directional differences in loss may be small but are real and are very hard to measure. They require very careful lab procedures to isolate other variables, not something that can be done with field OTDR or insertion loss testing. Remember that every OTDR test has light going through every fiber joint in both directions and, in multimode tests, with likely different mode fill conditions. So what the OTDR sees is some average value that may not correspond to actual loss in either direction.

When you do bidirectional OTDR testing, you measure the loss and backscatter effects in two directions and use some simple math (see Chapter 9 on OTDR testing) to remove the backscatter differences. While you can take out the backscatter differences with math, you end up averaging the loss from each direction, getting an average, not the actual loss.

OTDR Length Measurement Uncertainty

Many fiber optic measurements depend on length. Measuring the attenuation coefficient of fiber is one of them, involving both length and loss. The OTDR length measurement is made by measuring the round trip time for the test pulse to an event or the fiber end, then using the speed of light in the fiber to calculate length.

The index of refraction (IOR or N) of the glass in the core of the fiber is used to calculate the speed of light in the fiber. The speed of light in the fiber is the speed of light in a vacuum (C, 299,792 km/s) divided by the IOR. For a popular singlemode fiber, Corning SMF-28, the IOR at 1310 nm is 1.4675 (see table below), so the speed of light in the fiber is:

Speed = C/IOR = 299,792 km/s/1.4675 = 204,287 km/s

Knowing that speed, the OTDR measures the time and converts it into distance. The difference between in IOR most popular fibers is typically less than 1%. That difference is small enough to generally ignore unless you need to know the fiber or cable length more exactly.

But do not underestimate the difficulty of making this measurement. On a 1km cable, the OTDR will make a measurement of only 1/204,287 seconds or 4.895 microseconds times 2 for the round trip – less than 10 microseconds. A meter would be 1/1000 of that or 10 nanoseconds. That requires some sophisticated instrumentation.

OTDRs are often used to measure cable length and distance to a fault, which is especially important when finding the location of damage to a cable, often caused by visible accidents like dig-ups, but sometimes by hidden damage like rodent damage, cuts from directional boring, lightning strikes, or similar problems. In these cases, knowing the cable length is more important than the fiber length.

Most OSP loose tube cables have 1-2% excess fiber (less on ribbon cables) to prevent fiber stress under cable tension encountered in pulling or aerial installation. Some manufacturers of cable can provide the corrected index of refraction (IOR) or speed (group velocity) to use for the particular cable you are testing. If you do not know IOR or the ratio of excess fiber, you can estimate it or, if you have a long spool of cable, calibrate it as is discussed in Chapter 9.

Uncertainty In the Placement Of Markers
Another source of error when measuring length is the placement of markers. The placement of a marker can be affected by the shape of the trace where the marker is being placed, which can be a reflectance pulse if it is at a

connection or an open connector, a loss event without reflectance such as a fusion splice, or a fiber end which may or may not have reflectance.

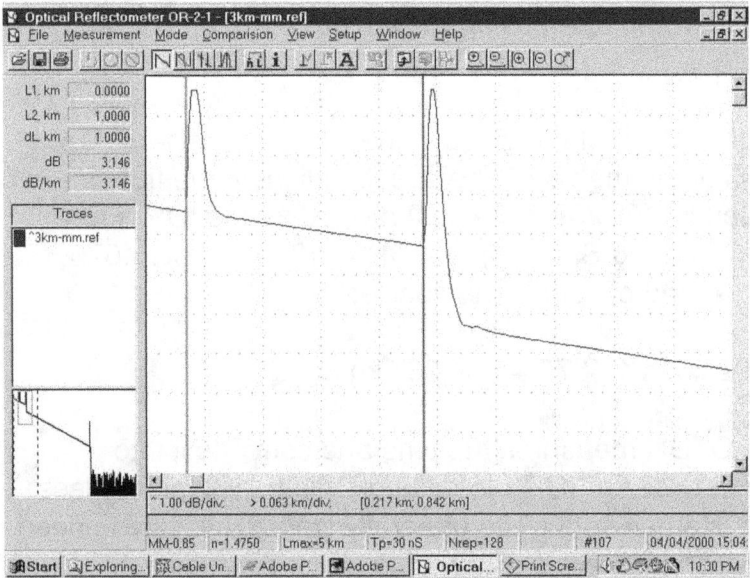

Manual placement of markers

When placing markers manually, it is important to be consistent. The markers should be placed just before the event. If the event is a connection and has a reflectance peak, the marker can easily be placed just at the point the peak rises from the backscatter level of the trace. If it is at a splice, place the marker just before where the splice loss drops from the backscatter level.

When measuring the length of a fiber or the distance to a break, finding the end of the fiber can be a problem. If the end of the fiber is cleaved or has a polished connector that has a high reflectance, it is easy to find from the strong reflectance peak.

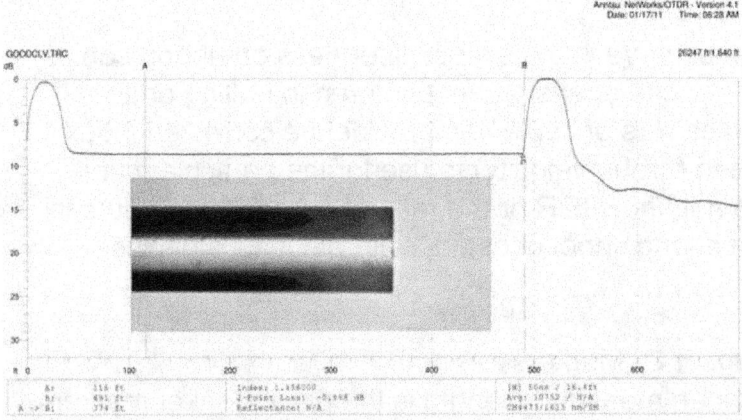

OTDR trace of cleaved fiber end
However a break will have a much smaller reflectance peak if any peak at all,

making the end of the fiber harder to locate.

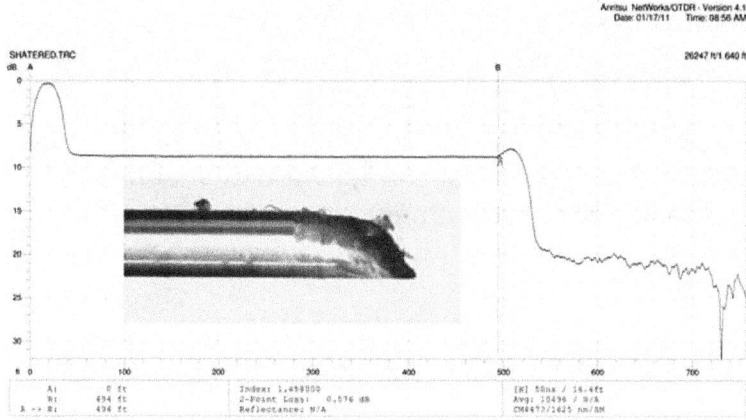

OTDR trace of broken fiber end

If the fiber is short and the end is broken, it's generally easy to see the fiber end and leave nothing but noise. On longer fibers that are near the limit of the OTDR, the trace itself may be noisy and the end hard to find. In that case, use a longer pulse width and/or more averaging to reduce the noise on the fiber so you can see the transition from backscatter to noise.

It can also be hard to place markers on low reflectance events like splices or connections, especially when the trace is noisy. As you can see in the trace below of an APC connector, the exact point where one places the marker is hard to determine. Think about the difficulty OTDR software in auto test would have with this trace.

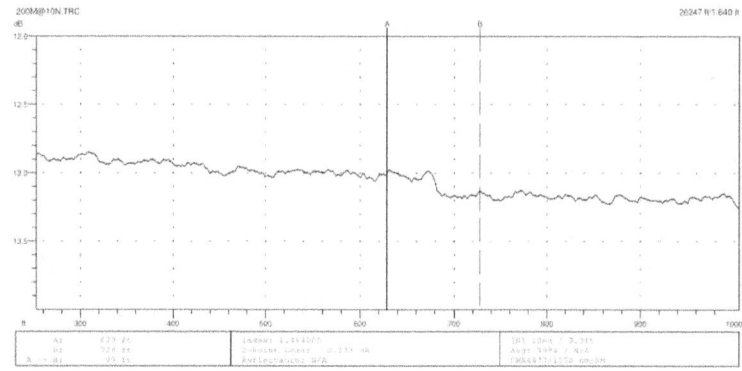

APC connector on noisy trace

Other Fiber Optic Measurements

There are many other fiber optic measurements that could be analyzed

similarly. For any measurement, the analysis involves understanding the measurement, understanding the variables that affect the measurement, knowing how the instruments and setups affect the measurements and creating a plan that minimized the negative effects on accuracy.

We have discussed the issues of measurement uncertainty in other fiber optic measurements included in this book and we encourage you to understand these principles and apply them to all measurements you make.

Chapter Exercises

- Test the same cable with both an OTDR and an OLTS and compare the results
- Make multiple measurements on the output of the source from initial turn on (cold) for 15-20 minutes and determine its warm-up time.
- Using the OLTS, test the same cable plant with a selection of launch and receive reference cables, record the data and calculate the average and standard deviation of the results.
- Test a cable plant with an OTDR using both manual marker settings and automatic testing and compare the results
- Test a connection or splice in a cable plant with an OTDR using both manual marker 2 point setting and LSA testing and compare the results.
- Using the OTDR, test the same connection or splice in a cable plant at least ten times, each time resetting the markers, record the data and calculate the average and standard deviation of the results.

Chapter Quiz

1. Scientists studying measurements at standards labs prefer the term "measurement uncertainty" to "accuracy."
>   True
>   False

2. Systematic errors mean measurements will all be in error but not all by the same amount.
>   True
>   False

3. In making fiber optic measurements, cleanliness of the connector can cause random errors.
>   True
>   False

4. The traditional mandrel wrap mode conditioner is a good approximation of the new encircled flux standard.
>   True
>   False

5. Bidirectional tests with an OTDR give the actual loss of a fiber joint, either a splice or a connection.
>   True
>   False

6. Fiber loss variations can be caused by test source wavelength and test source spectral width'
>   True
>   False

7. If a measurement had an average of -10.02 dB and as standard deviation of 0.20 dB, we could say the real measurement has a _____ likelihood of being in the range of -9.62 to -10.42 dB.
>   A. 33%
>   B. 68%
>   C. 95%
>   D. 100%

8. Which of the following can cause systematic errors in insertion loss testing?
>   A. Mode power distribution
>   B. Source wavelength
>   C. Dirt and contamination on connectors
>   D. Reference cable fiber core diameter

9. OTDRs show "gainers" because of _____.
>   A. Directional differences in backscatter
>   B. Different backscatter levels in different fibers
>   C. Wavelength shifts over long fiber runs
>   D. Pulse width variations due to fiber bandwidth

10. Encircled flux specifies that the power in the fiber is concentrated in the

_____.
>   A. Outside of the core
>   B. Center of the core
>   C. Inner 30% of the cladding
>   D. Joint between two fibers

11. The test to check source mode power distribution for encircled flux is called _____.
   A. EFF (EF factor)
   B. CPR (Coupled Power Ratio)
   C. HOML (Higher order mode loss)
   D. MPD (mode power distribution)

12. Multimode insertion loss tests with a controlled mode power distribution using encircled flux with a LED source will generally result in _____.
   A. Lower loss than without mode conditioning
   B. Higher loss than without mode conditioning
   C. About the same loss as without mode conditioning
   D. Close correlation to OTDR tests

Appendix A

References

There are other FOA materials that can be used as references for the testing and troubleshooting of a fiber optic cable plant or network. These include FOA printed textbooks and documents that can be accessed or downloaded from the FOA website.

FOA Textbooks
*   The FOA Reference Guide to Fiber Optics, by Jim Hayes, published by the FOA.
*   The FOA Reference Guide to Premises Cabling, by Jim Hayes, published by the FOA.
*   The FOA Reference Guide to Outside Plant Fiber Optics, by Jim Hayes, published by the FOA.
*   The FOA Reference Guide to Fiber Optic Network Design, by Jim Hayes, published by the FOA.

FOA Online Reference Guide, FOA website, www.foa.org

NECA/FOA-301 Standard For Installing And Testing Fiber Optic Cables
Download from FOA website

FOA Tech Bulletins (Printable Reference Documents)
*   Designing and manufacturing fiber optic communications products for manufacturers of products using fiber optics . (PDF, 0.2 Mb)
*   Choosing, installing and using fiber optic products for communications network users. (PDF, 0.1 Mb) (this document)
*   Designing Fiber Optic Networks - for contractors, designers, installers and users and the reference for the FOA CFOS/D Design Certification (PDF, 1.3 MB).
*   Installing Fiber Optic Cable Plants. (PDF, 0.2 Mb)
*   Troubleshooting fiber optic cable plants and communications systems. (PDF, 0.1 Mb)
*   Fiber Optic Restoration - how to plan ahead and restore networks quickly. (PDF, 0.1 Mb)

Contact The FOA
The Fiber Optic Association, Inc.
1119 S. Mission Road #355, Fallbrook, California 92028 USA
1-760-451-3655 Fax 1-781-207-2421
Email: info@foa.org   http://www.FOA.org

Appendix B

KSAs for FOA CFOS/T Specialist Certification In Testing

The ability to perform any job requires certain abilities, knowledge and skills, commonly referred to as "KSAs." For the fiber optic technician, these KSAs have been determined from more than 30 years of experience in actual installations. The FOA has developed this list as the requirements for CFOS/T specialist certification in fiber optic testing and to provide training organizations and instructors a list of topics that should be included in a training curriculum. For those working in the field who wish to become FOA certified, it is a list of relevant topics for study, whether using this textbook or the FOA Online Reference Guide.

Knowledge
This assumes the tech has the basic knowledge of a CFOT or equivalent.

Microscope inspection
  Types of microscopes
  Magnification for inspection for termination or cleanliness/condition

Cleaning Connectors
  How to inspect connectors for cleanliness
  Wet and Dry Cleaning and when to use each
  Procedures for wet and dry cleaning

Visual tracing and fault location
  Differences between fiber tracers and fault locators, where they are used and their limitations
  Use of fiber tracers, testing fiber polarity
  Use of fault locators, how to find basic faults

Optical Power Testing
  Measurement units of optical power
  Standards for optical power
  Features and options of fiber optic power meters
  Calibration of fiber optic power meters
  Power budgets for systems
  Measuring transmitter and receiver power
  Measurement uncertainty in optical power measurements

Loss Budgets

What is a loss budget, differences from power budget
Measurement units for optical loss
How to calculate loss budgets
Differences in loss budgets when testing with 1, 2 or 3 cable "0 dB" reference methods
Loss budgets for simple cable plants, PONs or networks with passive components

Insertion loss testing
How insertion loss simulates the actual operation of a cable plant
Basic instrument setup for insertion loss measurement
Measurement units for insertion loss measurements
Use of reference cables
How to set "0 dB" references for insertion loss testing
Importance of test source specifications
Mode power distribution requirements for multimode fiber testing
Mode power distribution requirements for singlemode fiber testing
Measurement uncertainty in insertion loss measurements
Calculating loss budgets for cables being tested, making pass/fail decisions

OTDR testing
How does OTDR testing work
How does OTDR testing differ from insertion loss testing
How does OTDR testing differ from the actual cable plant use
How to set up the OTDR for making measurements
Use of launch and receive cables
Meaning of the OTDR "dead zone"
How to measure length and loss by the 2-marker method
How to measure loss by the "least squares approximation" method
Why does one use the "least squares approximation" method
How does the OTDR measure reflectance
What are the error common to OTDR testing
Why do you make bi-directional measurements
Measurement uncertainty in OTDR measurements
Correllation of OTDR and insertion loss measurements

Fiber Characterization
Why does one test long haul SM cables for CD and PMD
What is CD, what causes it and how is it tested
What is PMD, what causes it and how is it tested
How does one compensate for CD and PMD
What is spectral attenuation and why is it important

Skills

The skills for testing specialist are similar to the CFOT skills for testing but specialists are expected to be skilled in setting up and using instruments for testing, maintaining their equipment, calculating loss budgets and using them to compare to actual test results.

Abilities

All CFOT abilities

Appendix C

Definitions of Terms

A

Absorption: That portion of fiber optic attenuation resulting of conversion of optical power to heat.

Analog: Signals that are continually changing, as opposed to being digitally encoded.

APC: Angled Physical Contact, APC Connector. A physical contact connector with an 8 degree angled end used on singlemode fiber to prevent reflectance.

Attenuation Coefficient: Characteristic of the attenuation of an optical fiber per unit length, in dB/km.

Attenuation: The reduction in optical power as it passes along a fiber, usually expressed in decibels (dB). See optical loss.

Attenuator: A device that reduces signal power in a fiber optic link by inducing loss.

Average power: The average over time of a modulated signal.

B

Back reflection, obsolete term, now called reflectance or optical return loss: Light reflected from the cleaved end of a fiber or a connection caused by the difference of refractive indices of air and glass. Expressed in dB relative to incident power.

Backscattering: The scattering of light in a fiber back toward the source, used to make OTDR measurements.

Bandwidth: The range of signal frequencies or bit rate within which a fiber optic component, link or network will operate.

Bending loss, microbending loss: Loss in fiber caused by stress on the fiber bent around a restrictive radius.

Bit-error rate (BER): The fraction of data bits transmitted that are received in error.

Bit: An electrical or optical pulse that carries information.

Buffer: A protective coating applied directly on the fiber.

C

Cable: One or more fibers enclosed in protective coverings and strength members.

Cable Plant, Fiber Optic: The combination of fiber optic cable sections, connectors and splices forming the optical path between two terminal devices.

CATV: An abbreviation for Community Antenna Television or cable TV.

CD, chromatic dispersion: The temporal spreading of a pulse in an optical waveguide caused by the wavelength dependence of the velocities of light in glass.

Cladding: The lower refractive index optical coating surrounding the core of a fiber that "traps" light in the core.

Connector: A device that provides for a demountable connection between two fibers or a fiber and an active device.

Core: The center of the optical fiber through which light is transmitted.

Coupled Power Ratio (CPR): an obsolete metric for measuring the modal distribution of a multimode LED test source.

Coupler: An optical device that splits or combines light from more than one fiber.

Cutback method: A technique for measuring the loss of bare fiber by measuring the optical power transmitted through a long length then cutting back to the source and measuring the initial coupled power.

Cutoff wavelength: The wavelength beyond which singlemode fiber only supports one mode of propagation.

CWDM: Coarse wavelength division multiplexing using lasers spaced widely over the range of 1260 to 1670 nm.

D

dBm: Optical power referenced to 1 milliwatt.

Decibel (dB): A unit of measurement of optical power which indicates relative power on a logarithmic scale, sometimes in the past called dBr. dB=10 log (power ratio)

Detector: A photodiode that converts optical signals to electrical signals.

DFB laser: Distributed feedback laser used for high speed and long distance transmitters.

Digital: Signals encoded into discrete bits.

Dispersion: The temporal spreading of a pulse in an optical waveguide. May be caused by modal chromatic or polarization effects.

E

EDFA: Erbium-doped fiber amplifier, an all optical amplifier for 1490-1650 nm SM transmission systems.

Edge-emitting diode (E-LED): A LED that emits from the edge of the semiconductor chip, producing higher power and narrower spectral width.

Encircled flux: A metric for the mode fill of multimode fiber used to specify test conditions.

End finish: The quality of the end surface of a fiber prepared for splicing or terminated in a connector.

Equilibrium modal distribution (EMD): Steady state modal distribution in multimode fiber, achieved some distance from the source, where the relative power in the modes becomes stable with increasing distance.

Excess loss: The amount of light lost in a coupler, beyond that inherent in the splitting to multiple output fibers.

F

Fiber Amplifier: an all optical amplifier using erbium or other doped fibers and pump lasers to increase signal output power without electronic conversion. See EDFA.

Ferrule: A precision tube which holds a fiber for alignment for interconnection or termination. A ferrule may be part of a connector or mechanical splice.

Fiber tracer: An instrument that couples visible light into the fiber to allow visual checking of continuity and tracing for correct connections.

Fiber identifier: A device that clamps onto a fiber and couples light from the fiber by bending, to identify the fiber and detect high speed traffic of an operating link or a 2 kHz tone injected by a test source.

Fiber optics: Light transmission through flexible optical fibers for communications or lighting.

FO: Common abbreviation for "fiber optic."

Fresnel reflection, back reflection, optical return loss: Light reflected from the cleaved or polished end of a fiber caused by the difference of refractive indices of air and glass. Typically 4% of the incident light for an air-glass interface.

FTTH: fiber to the home

Fusion splicer: An instrument that splices fibers by fusing or welding them, typically by electrical arc.

G

Graded index (GI): A type of multimode fiber which used a graded profile of refractive index in the core material to correct for dispersion.

I

Index of refraction: A measure of the speed of light in a material.

Index matching fluid: A liquid used of refractive index similar to glass used to match the materials at the ends of two fibers to reduce loss and back reflection.

Index profile: The refractive index of a fiber as a function of cross section.

Insertion loss: The loss caused by the insertion of a component such as a splice or connector in an optical fiber.

J

Jacket: The protective outer coating of the cable.

Jumper cable: A short single fiber cable with connectors on both ends used for interconnecting other cables or testing.

L

Laser diode, ILD: A semiconductor device that emits high powered, coherent light when stimulated by an electrical current. Used in transmitters for singlemode fiber links.

Launch cable: A known good fiber optic jumper cable attached to a source and calibrated for output power used used as a reference cable for loss testing. This cable must be made of fiber and connectors of a matching type to the cables to be tested.

Light-emitting diode, LED: A semiconductor device that emits light when stimulated by an electrical current. Used in transmitters for multimode fiber links.

Link, fiber optic: A combination of transmitter, receiver and fiber optic cable connecting them capable of transmitting data. May be analog or digital.

Long wavelength: A commonly used term for light in the 1300 and 1550 nm ranges.

Loss, optical: The amount of optical power lost as light is transmitted through fiber, splices, couplers, etc.

Loss budget: The amount of power lost in the link. Often used in terms of the maximum amount of loss that can be tolerated by a given link.

M

Margin: The additional amount of loss that can be tolerated in a link.

Mechanical splice: A semi-permanent connection between two fibers made with an alignment device and index matching fluid or adhesive.

Micron (*m): A unit of measure, 10-6 m, used to measure wavelength of light.

Microscope, fiber optic inspection: A microscope used to inspect the end surface of a connector for flaws or contamination or a fiber for cleave quality.

Modal dispersion: The temporal spreading of a pulse in an optical waveguide caused by modal effects.

Mode field diameter: A measure of the core size in singlemode fiber.

Mode filter: A device that removes optical power in higher order modes in fiber.

Mode scrambler: A device that mixes optical power in fiber to achieve equal power distribution in all modes. Mode stripper: A device that removes light in the cladding of an optical fiber.

Mode: A single electromagnetic field pattern that travels in fiber.

Multimode fiber: A fiber with core diameter much larger than the wavelength of light transmitted that allows many modes of light to propagate. Commonly used with LED sources for lower speed, short distance links.

N

Nanometer (nm): A unit of measure , 10-9 m, used to measure the wavelength of light.

Network: A system of cables, hardware and equipment used for communications.

Numerical aperture (NA): A measure of the light acceptance angle of the fiber.

O

Optical amplifier: A device that amplifies light without converting it to an electrical signal.

Optical fiber: An optical waveguide, comprised of a light carrying core and cladding which traps light in the core.

Optical loss test set (OLTS): An measurement instrument for optical loss that includes both a meter and source.

Optical power: The amount of radiant energy per unit time, expressed in linear units of Watts or on a logarithmic scale, in dBm (where 0 dB = 1 mW) or dB* (where 0 dB*=1 microWatt).

Optical switch: A device that routes an optical signal from one or more input ports to one or more output ports.

Optical time domain reflectometer (OTDR): An instrument that used backscattered light to measure fiber parameters and find faults.

ORL, optical return loss, reflectance, back reflection: Light reflected from the cleaved or polished end of a fiber or a connection caused by the difference of refractive indices of air and glass. For cable plants, may also include fiber backscatter. Expressed in dB relative to incident power.

Overfilled launch: A condition for launching light into the fiber where the incoming light has a spot size and NA larger than accepted by the fiber, filling all modes in the fiber.

P

PC: Physical Contact, PC Connector: A cpnnectow with a convex ferrule intended to ensure fiber to fiber contact for lower reflectance.

Photodiode: A semiconductor that converts light to an electrical signal, used in fiber optic receivers.

Pigtail: A short length of fiber attached to a fiber optic component such as a laser or coupler.

Plastic optical fiber (POF): An optical fiber made of plastic.

Plastic-clad silica (PCS) fiber: A fiber made with a glass core and plastic cladding.

Polarization mode dispersion (PMD): Dispersion in singlemode fiber caused by the difference in speed of light of the polarization modes in the fiber.

Power budget: The difference (in dB) between the transmitted optical power (in dBm) and the receiver sensitivity (in dBm).

Power meter, fiber optic: An instrument that measures optical power emanating form the end of a fiber.

Preform: The large diameter glass rod from which fiber is drawn.

Prepolished/splice connector: A factory-made connector with a fiber stub that is spliced onto fiber for termination. Also called SOC, splice on connector.

R

Receive cable: A known good fiber optic jumper cable attached to a power meter used as a reference cable for loss testing. This cable must be made of fiber and connectors of a matching type to the cables to be tested.

Receiver: A device containing a photodiode and signal conditioning circuitry that converts light to an electrical signal in fiber optic links.

Reference cable: A known good fiber optic jumper cable attached to a light source or power meter used as a reference cable for loss testing.

Reflectance: Light reflected from the cleaved or polished end of a fiber caused by the difference of refractive indices of air and glass.

Refractive index: A property of optical materials that relates to the velocity of light in the material.

Repeater, regenerator: A device that receives a fiber optic signal and regenerates it for retransmission, used in very long fiber optic links.

S

Scattering: The change of direction of light after striking small particles that causes loss in optical fibers.

Sheath: The term used for the outer protective layers of a cable consisting of jacket, armor and strength members.

Short wavelength: A commonly used term for light in the 665, 790, and 850 nm ranges.

Singlemode fiber: A fiber with a small core, only a few times the wavelength of light transmitted, that only allows one mode of light to propagate. Commonly used with laser sources for high speed, long distance links.

Source: A laser diode or LED used to inject an optical signal into fiber.

Spectral Attenuation (SA): the attenuation of an optical fiber over the full range of wavelengths it may be used, e.g. with wavelength division multiplexing. Sometimes also called Attenuation Profiling.

Splice (fusion or mechanical): A device that provides for a connection between two fibers, typically intended to be permanent.

Splice On Connector (SOC): A factory-made connector with a fiber stub which is spliced onto fiber for termination. Also called prepolished/splice connector.

Splitting ratio: The distribution of power among the output fibers of a coupler.

Steady state modal distribution: Equilibrium modal distribution (EMD) in multimode fiber, achieved some distance from the source, where the relative power in the modes becomes stable with increasing distance.

Step index fiber: A multimode fiber where the core is all the same index of refraction.

Surface emitter LED: A LED that emits light perpendicular to the semiconductor chip. Most LEDs used in datacommunications are surface emitters.

T

Talkset, fiber optic: A communication device that allows conversation over unused fibers.

Termination: Preparation of the end of a fiber to allow connection to another fiber or an active device, sometimes also called "connectorization".

Test cable: A short single fiber jumper cable with connectors on both ends used for testing. This cable must be made of fiber and connectors of a matching type to the cables to be tested.

Test kit: A kit of fiber optic instruments, typically including a power meter, source and test accessories used for measuring loss and power.

Test source: A laser diode or LED used to inject an optical signal into fiber for testing loss of the fiber or other components.

Total internal reflection: Confinement of light into the core of a fiber by the reflection off the core-cladding boundary.

Transmitter: A device which includes a LED or laser source and signal conditioning electronics that is used to inject a signal into fiber.

V

VCSEL: vertical cavity surface emitting laser, a type of laser that emits light vertically out of the chip, not out the edge, widely used in fast multimode networks.

Visual fault locator (VFL): A device that couples visible light into the fiber to allow visual tracing and testing of continuity. Some are bright enough to allow finding breaks in fiber through the cable jacket.

W

Watts: A linear measure of optical power, usually expressed in milliwatts (mW), microwatts (*W) or nanowatts (nW).

Wavelength: A measure of the color of light, usually expressed in nanometers (nm) or microns (*m).

Wavelength division multiplexing (WDM): A technique of sending signals of several different wavelengths of light into the fiber simultaneously.

Working margin: The difference (in dB) between the power budget and the loss budget (i.e. the excess power margin).

Index

Finding Things In This Book

This book is adopting a lot of new ideas in creating a more useful book for reference and training, so we thought we'd try a new approach to the index also. Often when trying to find something in an index, you end up looking at dozens of pages before you find what you want.

We're going to try another approach, closer to a detailed Table of Contents with comments, organized by topics we think will be most likely sought. Start with the area of interest, then look for the subjects below. Let us know what you think.

The FOA Online Reference Guide has a Google Custom Search function that can help find specific topics or terms on the FOA site.

If you are looking for a definition of a term used in Premises Cabling, start here. If you are looking for a specific definition of a technical word, there is a Glossary in the FOA Online Reference Guide.

Chapter 9
Optical Time Domain Reflectometer (OTDR) Testing

Chapter 14
Metrology and Fiber Optic Measurement Uncertainty 267

www.ingramcontent.com/pod-product-compliance
Lightning Source LLC
Chambersburg PA
CBHW080235180526
45167CB00006B/2281